HOW TO USE A SCIENTIFIC CALCULATOR

The Essential Companion for
Students and Professionals

Fitzpatrick J. Thompkins

Copyright © 2024 by **Fitzpatrick J. Thompkins**

All rights reserved

No part of this publication may be reproduced, stored in a retrieval system, or transmitted, in any form or by any means, electronic, mechanical, photocopying, recording, or otherwise, without the prior written permission of the author.

The information in this ebook is true and complete to the best of our knowledge. All recommendation are made without guarantee on the part of author or publisher. The author and publisher disclaim any liability in connection with the use of this information.

Table of Contents

Introduction   **6**
    Overview of the Guide   9
    Importance of Learning to Use a Scientific Calculator   12
    Different Brands and Models of Scientific Calculators   15

Chapter: 1 Getting Started   **18**
    Unboxing and Initial Setup   18
    Powering Your Calculator On and Off   21
    Basic Navigation and Key Functions   24
    Customizing Settings for Personal Preferences   27

Chapter: 2 Understanding the Basics   **31**
    Arithmetic Operations: Addition, Subtraction, Multiplication, and Division   31
    Using Memory Functions   35
    Working with Percentages and Fractions   39
    Introduction to Scientific Notation   43

Chapter: 3 Advanced Mathematical Functions   **46**
    Trigonometric Functions: Sine, Cosine, Tangent, and Their Inverses   46
    Logarithmic and Exponential Functions   50
    Calculating Derivatives and Integrals   54
    Complex Numbers: Basic Operations and Polar Form   58

Chapter: 4 Programming and Custom Functions   **62**
    Introduction to Calculator Programming   62

Creating and Storing Custom Functions	66
Using Pre-installed Applications	70
Chapter: 5 Statistical Functions	**74**
Entering and Editing Data Sets	74
Calculating Mean, Median, Mode, and Standard Deviation	78
Regression Analysis and Plotting Graphs	81
Chapter: 6 Graphing Capabilities	**85**
Understanding the Graphing Interface	85
Plotting and Analyzing Functions	89
Using Zoom and Trace Features	93
Saving and Exporting Graphs	97
Chapter: 7 Troubleshooting and Maintenance	**101**
Common Issues and How to Solve Them	101
Updating Calculator Firmware	105
Cleaning and Caring for Your Calculator	109
Chapter: 8 Tips and Tricks	**113**
Shortcuts for Efficient Calculation	113
Recommended Resources for Learning	117
Practice Problems to Enhance Skills	121
Chapter: 9 Comparing Scientific Calculators	**125**
Features to Consider When Buying a Calculator	125
Pros and Cons of Popular Models	129
Recommendations Based on Use Case	133
Chapter: 10 Ethical Considerations and Exam Policies	**137**
Using Calculators Responsibly	137
Understanding Examination Rules Regarding Calculators	140

Conclusion 144

Introduction

In a busy town, there was a special bookstore known for its rare educational books. One of them, "How to Use a Scientific Calculator," became legendary among students and pros.

The story starts with Sam, a freshman in engineering who struggled with math. During a tough exam, Sam's old calculator couldn't keep up, leading to a bad grade.

Determined to improve, Sam visited the bookstore. The owner suggested "How to Use a Scientific Calculator," promising it would help understand math better.

Skeptical but curious, Sam started reading. The book taught not just how to use a calculator but also the reasons behind each function, linking math to real-life stories and applications.

With practice, Sam's math skills improved dramatically. Sam's grades soared, and they became more confident in class discussions.

The book ended with a lesson: learning isn't just about grades, but about thinking critically and solving problems creatively.

Sam's story became a testament to the book's power, showing how it could transform anyone's approach to math with curiosity and confidence.

Overview of the Guide

This guide is crafted with the intention of demystifying the complexities and harnessing the full potential of the scientific calculator, a tool indispensable for students, professionals, and anyone engaged in the fields of science, technology, engineering, and mathematics. It aims to transition its readers from novices to proficient users by covering a wide array of functionalities, from basic arithmetic operations to more sophisticated features like statistical analysis, graphing capabilities, and even programming.

The journey begins by familiarizing readers with the initial setup and basic navigation of their scientific calculators, setting a solid foundation on which to build more advanced skills. As the guide progresses, it delves into arithmetic operations and the use of memory functions, ensuring that users can efficiently perform calculations and store results for future reference. The exploration of percentages, fractions, and scientific notation further enhances the reader's ability to tackle a broad spectrum of mathematical problems with ease.

Moving beyond the basics, the guide introduces trigonometric functions, logarithms, and exponential functions, demystifying these often intimidating areas of mathematics. It explains how to calculate derivatives and integrals, offering insights into the calculator's utility for higher-level math courses and professional applications. The section on complex numbers not only covers basic operations but also teaches readers how to utilize the polar form for more complex calculations.

The guide recognizes the importance of customization and programming for users who wish to tailor their calculators to their specific needs. It provides step-by-step instructions on creating and storing custom functions, as well as utilizing pre-installed applications to maximize efficiency and productivity. This aspect of the guide is particularly beneficial for users who engage in repetitive calculations or those who require specialized functions for their studies or professional work.

Statistical functions receive thorough coverage, equipping readers with the skills to enter and edit data sets, calculate descriptive statistics, and perform regression analysis. The guide emphasizes the calculator's graphing capabilities, explaining how to plot functions, analyze graphs, and use features like zoom and trace to gain deeper insights into mathematical relationships.

Understanding that even the most advanced users may encounter challenges, the guide includes a troubleshooting section. This part addresses common issues, provides solutions, and offers tips for maintenance to ensure the calculator remains a reliable tool for years to come. Additionally, it presents shortcuts and tips for efficient calculation, along with recommended resources for those who wish to further their understanding.

In concluding, the guide goes beyond the technical aspects of using a scientific calculator. It touches upon ethical considerations and exam policies, ensuring readers are well-informed about responsible usage, especially in academic settings. By weaving together technical instructions, practical applications, and engaging narrative elements, this guide transforms the daunting task of mastering a scientific calculator into an achievable and rewarding endeavor. It encourages readers to view their calculator not just as a tool for calculation, but as a gateway to deeper understanding and innovation in their academic and professional pursuits.

Importance of Learning to Use a Scientific Calculator

The importance of learning to use a scientific calculator transcends the mere operation of a device; it represents a fundamental skill set in the modern educational and professional landscape. This device, compact yet powerful, is a gateway to understanding and applying complex mathematical concepts across various fields, including science, engineering, finance, and beyond. Mastering its use can significantly enhance one's ability to solve problems efficiently and accurately, fostering a deeper comprehension of the material at hand.

A scientific calculator offers functionalities that go far beyond basic arithmetic operations. It includes the capability to perform operations involving exponents, logarithms, trigonometric functions, and more, enabling users to tackle a wide range of mathematical problems. The nuanced understanding required to navigate these functions encourages users to engage with mathematical concepts on a more intimate level, thus deepening their grasp of the subject matter. As such, proficiency with a scientific calculator becomes not just a matter of convenience but a critical component of academic and professional success.

Furthermore, learning to use a scientific calculator effectively can significantly improve one's problem-solving skills. It teaches users to break down complex problems into simpler, more manageable parts, identifying which functions and operations to apply and in what sequence. This process cultivates logical thinking and precision, skills that are invaluable across disciplines and in daily life. It also encourages the exploration of alternative solutions to a problem, fostering creativity and innovation.

In the context of education, the ability to use a scientific calculator efficiently can be particularly empowering for students. It levels the playing field, ensuring that students focus more on understanding concepts rather than getting bogged down by lengthy calculations. This efficiency can lead to better performance on exams and assignments, where time management is crucial. Moreover, in many standardized tests and certification exams, a scientific calculator is an allowed, if not necessary, tool, making its mastery a non-negotiable skill.

The relevance of scientific calculators extends into the professional world as well. In fields such as engineering, physics, and finance, the ability to quickly perform complex calculations can significantly enhance productivity and accuracy. It enables professionals to make informed decisions based on precise data analysis, a cornerstone of quality work in these and many other fields. Thus, proficiency with a scientific calculator is not just about performing mathematical operations; it's about applying these operations to solve real-world problems effectively.

Equally important is the role of scientific calculators in fostering a lifelong interest in mathematics and science. By demystifying complex calculations and making them more accessible, these devices can help remove barriers to learning and exploration in these subjects. For many students, the ability to easily navigate mathematical challenges with the help of a scientific calculator can be the key to developing a passion for STEM fields, potentially guiding their educational and career paths.

In summary, learning to use a scientific calculator is a critical skill that offers wide-ranging benefits, from enhancing academic performance and problem-solving skills to supporting professional success and fostering a deeper interest in STEM fields. It equips individuals with the tools to not only tackle complex mathematical challenges but also to approach problems in a logical, precise, and creative manner. As such, the role of the scientific calculator in education and professional development cannot be overstated, making it an indispensable tool in the modern world.

Different Brands and Models of Scientific Calculators

When exploring the diverse world of scientific calculators, it becomes apparent that each brand and model brings its unique features, functionalities, and user interfaces to the table, catering to a wide range of needs from high school students to professionals in engineering and the sciences. The landscape of these calculators is dominated by several key players, including but not limited to Texas Instruments, Casio, and Hewlett-Packard, each offering a variety of models that have evolved over time to incorporate more advanced computing capabilities and user-friendly features.

Texas Instruments, often abbreviated as TI, is renowned for its robust range of scientific calculators, with the TI-30X and TI-36X Pro models being particularly popular among students for their affordability and ease of use. These calculators are designed with education in mind, offering features that support classroom learning and standardized testing. Moving up the range, Texas Instruments introduces graphing capabilities with models like the TI-84 Plus and TI-Nspire CX, which are staples in high school and college mathematics courses. The TI-Nspire CX, especially, stands out for its ability to perform a wide array of functions, from algebraic computations to calculus, not to mention its rechargeable battery and color display which enhance its usability.

Casio, another heavyweight in the scientific calculator market, competes closely with Texas Instruments. The Casio FX series, particularly the FX-115ES Plus and FX-991EX (ClassWiz), are celebrated for their high-resolution displays and extensive functionality that spans from basic scientific calculations to more complex equations involving calculus and statistics. The ClassWiz series, for example, boasts a natural textbook display that allows users to enter and view mathematical expressions as they would appear in a textbook, making it easier for users to learn and perform complex calculations. Casio also dips into the graphing calculator market with models like the FX-CG50 (Prizm), which provides a color display and the ability to graph in three dimensions, a feature that appeals to both students and professionals who require detailed graphical analyses.

Hewlett-Packard, or HP, offers a range of scientific calculators that appeal to professionals and enthusiasts who prefer RPN (Reverse Polish Notation), a feature unique to HP calculators. The HP 35s and HP Prime are notable models in their lineup, with the HP Prime being particularly advanced, featuring a touchscreen interface, wireless connectivity, and a robust computing engine capable of handling a wide range of mathematical and scientific applications. The HP Prime caters to a higher education and professional market, offering functionalities that are beneficial for engineers, mathematicians, and scientists.

Beyond these leading brands, there are other manufacturers such as Sharp and TI, which also offer competitive models, though with less market presence. Each model, regardless of brand, has been designed with specific user needs in mind, from the basic scientific calculators suitable for middle and high school students to more advanced models that support higher education and professional work in fields such as engineering, physics, and mathematics.

Understanding the functionalities, strengths, and limitations of these various brands and models is crucial for anyone looking to maximize their use of a scientific calculator. Whether it's for solving simple arithmetic operations, performing complex calculus functions, or even programming custom applications, there exists a scientific calculator designed to meet those specific needs. This variety ensures that learners and professionals can select a device that not only fits their computational requirements but also aligns with their personal preferences for interface design, display type, and input method, thereby enhancing their overall learning and working experience.

Chapter: 1 Getting Started

Unboxing and Initial Setup

The process of unboxing and setting up a new scientific calculator is the first step in a journey toward mastering an essential tool for academic and professional success in fields that require complex mathematical computations. This initial phase is not only about physical setup but also about familiarizing oneself with the device's features, capabilities, and personalization options. Let's walk through a comprehensive overview of what this entails, emphasizing the universal aspects applicable to most scientific calculators while acknowledging the slight variations across different brands and models.

Upon opening the box, the first item most users encounter is the calculator itself, nestled securely within protective packaging to prevent damage during transit. Accompanying the calculator, one typically finds a user manual, which is an invaluable resource for understanding the device's functions and features. Some models might also include a protective case or slide cover to safeguard the calculator against drops and scratches, as well as batteries or a charging cable, depending on the power requirements of the device.

The initial setup begins with powering on the device. For calculators that operate on batteries, this may involve installing the batteries provided. In the case of solar-powered models or those with a rechargeable battery, ensuring the device is sufficiently charged or exposed to light is crucial. Once powered on, the calculator typically displays a greeting or initialization screen, indicating it is ready for use.

The next step is to familiarize oneself with the calculator's basic layout. This includes identifying the primary keys for arithmetic operations, the on/off button, the mode settings, and any special function keys. The user manual can be particularly helpful in this regard, offering detailed explanations of each key's purpose and how to access the calculator's various functions.

Configuring the device settings is an important aspect of the initial setup. Most scientific calculators allow users to customize operational modes, such as degree or radian for angle measurements, floating or fixed decimal places for display, and other preferences that affect how calculations are performed and results are displayed. Accessing these settings typically involves pressing a specific combination of keys, as outlined in the user manual.

For calculators with advanced features, such as graphing capabilities or programmable functions, the initial setup may also include a brief exploration of these options to understand how they are accessed and used. While a detailed understanding of these features may develop over time, getting a sense of what the calculator can do beyond basic arithmetic operations is beneficial.

Finally, the setup process should include a test run of basic calculations to ensure the device is working correctly. This can involve performing simple arithmetic operations, using the scientific notation feature, or exploring the trigonometric functions. Through this hands-on experience, users begin to develop a feel for the calculator's interface and responsiveness, setting the stage for more advanced usage.

The unboxing and initial setup of a scientific calculator, while straightforward, are critical steps in becoming comfortable and proficient with this powerful tool. By taking the time to properly install batteries or charge the device, explore its layout and settings, and conduct initial test calculations, users can ensure they are prepared to tackle a wide range of mathematical challenges. This foundation also enables users to more effectively dive into the deeper functionalities and customizations that their specific model of calculator offers, maximizing the device's potential as a partner in their academic and professional endeavors.

Powering Your Calculator On and Off

Mastering the operation of a scientific calculator begins with understanding the fundamental steps of powering the device on and off, a process that, while seemingly straightforward, is essential for efficient and effective use. Each calculator brand and model may have its unique method for performing these basic operations, yet there are commonalities across devices that users can familiarize themselves with to ensure they're getting started on the right foot.

Powering on a scientific calculator typically involves pressing a clearly marked "On" or "Power" button, often located in the upper section of the keypad. This action brings the calculator to life, illuminating its display and readying it for input. In some models, particularly those from Texas Instruments or Casio, the "On" button might serve dual purposes, also functioning as a wake-up call for calculators in sleep mode, a power-saving state where the calculator remains on but uses minimal energy. For devices without a specific "On" button, pressing any key might suffice to activate the calculator from its dormant state.

Turning off a scientific calculator can be slightly less intuitive, given that some models do not feature a dedicated "Off" button. In such cases, users might need to execute a specific key combination or hold down the

"On" button for a certain duration to power down the device. For example, Texas Instruments calculators often require the user to press a sequence like "2nd" followed by "Off" to shut down the calculator. Casio models, on the other hand, might automatically enter a sleep mode after a period of inactivity, conserving battery life without fully turning off.

Understanding the power management features of a scientific calculator is crucial, as it not only facilitates the calculator's readiness for use but also contributes to the longevity of the device. Many scientific calculators are equipped with solar panels in addition to battery compartments, harnessing natural or artificial light to extend battery life. For these dual-powered calculators, even in off mode, placing them in well-lit areas can keep the internal battery charged and prolong the intervals between battery replacements.

Battery management is another aspect of powering your calculator that cannot be overlooked. Users should be familiar with their calculator's battery type (commonly AAA or button cell batteries) and know how to safely replace them when necessary. Some advanced models will have low-battery indicators or messages on the display, prompting users to change the batteries soon to avoid losing important data or custom settings.

In the realm of digital convenience and sustainability, some of the latest models of scientific calculators come with rechargeable batteries. These calculators are often equipped with a USB port, allowing for easy charging via a computer or power adapter. Users of these models need to monitor the battery level indicator and ensure their device is charged regularly, especially before exams or critical use periods, to guarantee uninterrupted operation.

Powering your calculator on and off may seem like the most basic of tasks, yet it embodies the importance of familiarity with your device for seamless and efficient use. By understanding these initial steps, along with the accompanying power management strategies, users can ensure their scientific calculators serve them well throughout their academic and professional endeavors, always ready at a moment's notice to tackle the next complex problem.

Basic Navigation and Key Functions

Mastering the basic navigation and key functions of a scientific calculator is the first step in unlocking its full potential. These devices, designed to perform operations ranging from simple arithmetic to complex scientific equations, have a standardized yet unique set of keys and navigation methods across different brands and models. Understanding these basics not only enhances computational efficiency but also lays the foundation for exploring more advanced features.

At the heart of any scientific calculator is the keypad, divided into several sections, each dedicated to a specific type of function. The most common layout includes a numerical keypad, function keys for operations like addition, subtraction, multiplication, and division, and a cluster of scientific function keys that allow for computations involving exponents, roots, trigonometry, and more. Additionally, most calculators feature a mode key, which toggles between different operational modes such as scientific, statistical, and programming, allowing the device to adapt to various mathematical tasks.

One of the first skills to master is entering numbers and performing basic arithmetic operations. This involves using the numerical keypad and the basic operation keys. Calculators typically follow the standard order of operations, automatically prioritizing multiplication and division over addition and subtraction, unless parentheses are used to specify a different order. This intuitive design ensures that users can input equations as they would be written on paper.

Scientific functions are accessed either directly through dedicated keys or through a shift or second function key, which unlocks a secondary set of operations labeled on the calculator. This dual-functionality design maximizes the calculator's capabilities without overwhelming the user with an excessive number of keys. For instance, pressing the shift key followed by the sine key might access the inverse sine function, allowing for a wide range of trigonometric calculations without necessitating additional keys.

Memory functions play a crucial role in managing calculations. The most common memory operations include storing a number in memory, recalling a stored number, adding to or subtracting from a stored number, and clearing the memory. These functions are indispensable for complex calculations where intermediate results need to be saved for later use.

Navigating through the calculator's menu and settings is another fundamental aspect. Most scientific calculators feature a menu button that provides access to various settings, such as angle measurement units (degrees, radians, grads), display format (normal, scientific notation, fixed decimal), and other calculator-specific settings. Learning to efficiently navigate through these menus and settings allows users to customize their calculators according to their preferences and the requirements of specific calculations.

Clearing entries and errors is also crucial. Calculators generally have a clear or delete key that allows users to correct mistakes or clear the current operation without resetting the entire device. This function is especially useful in a learning environment, where trial and error form an essential part of the learning process.

Understanding the basic navigation and key functions of a scientific calculator requires patience and practice. Familiarity with these fundamentals not only makes mathematical calculations more manageable but also builds confidence in using the calculator as a powerful tool for learning and problem-solving. As users become more comfortable with these basics, they can begin to explore the more advanced features of their calculators, further expanding their mathematical capabilities.

Customizing Settings for Personal Preferences

Customizing the settings of a scientific calculator to align with personal preferences plays a crucial role in enhancing the efficiency and comfort of its use. This customization allows users to adjust their calculators based on their own learning styles, professional needs, or simply their preferences for how information is displayed and interacted with. Understanding how to tailor these settings can transform the calculator from a mere tool into a personalized instrument that significantly aids in mathematical and scientific explorations.

Initially, one of the most fundamental settings that users might adjust is the display contrast and brightness. This adjustment is particularly important for ensuring that the calculator's screen is readable in a variety of lighting conditions, from the brightly lit classrooms and offices to the dimmer environments of early mornings or late nights spent studying. Users can typically find these settings in a setup or settings menu, allowing for adjustments to be made so that the display is comfortable for the eyes over extended periods.

Another important aspect of customization involves choosing the display mode for numbers. Scientific calculators often allow users to switch between standard, scientific, and engineering notations. This flexibility is vital for users across different fields; for example, engineers might prefer engineering notation, while those in pure sciences may opt for scientific notation. Additionally, the ability to define the number of decimal places or significant figures displayed in calculations is invaluable for precision in professional settings or when following specific academic guidelines.

For users who frequently deal with particular types of calculations, customizing the calculator's function keys can lead to a significant increase in efficiency. Some advanced models of scientific calculators allow users to assign frequently used operations or complex formulas to specific keys. This feature is a tremendous time-saver, reducing the need to repeatedly input complex sequences of operations.

Customization also extends to the calculator's operational mode, such as choosing between RPN (Reverse Polish Notation) and algebraic modes of input, depending on the user's preference or professional requirements. RPN, favored for its efficiency in certain calculations, may not be the default mode on many calculators but can usually be enabled through the settings. Understanding and selecting an input mode that aligns with one's thought process can significantly streamline calculations.

Moreover, for educators or professionals working in multicultural environments, the ability to switch the language of the calculator's interface is an often-overlooked aspect of customization. This setting, where available, can make the calculator more accessible and user-friendly, removing language barriers to understanding and using the device effectively.

The process of personalizing a calculator may also involve adjusting sound settings, where users can turn click sounds on or off, based on their preference for auditory feedback during key presses. While some find the feedback reassuring, others may prefer a silent operation, especially in quiet study environments or during examinations.

For those who delve deeper into the capabilities of their calculators, like those with graphing functions, customizing the appearance of graphs—including color schemes, grid styles, and default window settings—can make data visualization more intuitive and tailored to specific types of data analysis.

Incorporating personalized settings into the use of a scientific calculator can profoundly impact its utility and effectiveness as a tool for learning and professional work. It allows the device to become more than just a standard tool, transforming it into a tailored instrument finely tuned to the user's specific needs and preferences. This level of customization ensures that the calculator is not only a companion in solving mathematical problems but also a reflection of the user's personal approach to learning and exploration in the realms of science and mathematics.

Chapter: 2 Understanding the Basics

Arithmetic Operations: Addition, Subtraction, Multiplication, and Division

Understanding arithmetic operations is foundational to both mathematics and everyday problem-solving. These operations include addition, subtraction, multiplication, and division. A scientific calculator, a tool designed to perform not only basic arithmetic but also complex mathematical functions, can greatly aid in these computations. Here, we delve into the basics of arithmetic operations and elucidate how to utilize a scientific calculator for these purposes.

Addition

Definition: Addition is the process of combining two or more numbers to get a total or sum.

On a Scientific Calculator: To perform addition on a scientific calculator, input the first number, press the "+" (plus) button, enter the second number, and then press the "=" (equals) button to obtain the result. For adding more than two numbers, continue to press the "+" button followed by the next number, and hit "=" once all numbers are entered.

Subtraction

Definition: Subtraction involves taking one number away from another to find the difference between them.

On a Scientific Calculator: Start with entering the number you are subtracting from (minuend), press the "-" (minus) button, input the number you are subtracting (subtrahend), and press "=" to see the difference. If performing multiple subtractions in sequence, continue pressing the "-" button followed by the numbers you wish to subtract, hitting "=" at the end.

Multiplication

Definition: Multiplication is the process of adding a number to itself a certain number of times, which is equivalent to the repeated addition of the same number.

On a Scientific Calculator: Enter the first number, press the "×" (multiply) button, enter the second number, and press "=" to find the product. For multiplying several numbers, after pressing "=", continue by pressing the "×"

button, entering the next number, and pressing "=" again after the last number.

Division

Definition: Division is the operation of distributing a number (dividend) into equal parts, determining how many times one number is contained within another.

On a Scientific Calculator: Input the dividend (the number you're dividing), press the "÷" (divide) button, enter the divisor (the number by which you're dividing), and press "=" to obtain the quotient. For sequential division, after pressing "=", continue with the "÷" button, enter the next divisor, and press "=" after the last divisor.

Tips for Using a Scientific Calculator:

Check Mode: Ensure your calculator is in the correct mode (e.g., degree for trigonometric operations, general for basic arithmetic).
Clearing Entries: Use the "C" or "CE" button to clear your current entry if you make a mistake.
Parentheses for Order of Operations: Use parentheses "(" and ")" to specify the order of operations, particularly for complex calculations combining multiple operations.

Memory Functions: Utilize the memory functions (M+, M-, MR, MC) to store and recall numbers during lengthy calculations.

Error Checking: If the calculator displays an error, double-check your entries for any possible input mistakes.

Understanding these basic arithmetic operations and how to perform them using a scientific calculator can significantly enhance your efficiency in solving mathematical problems. Whether you're working on homework, preparing for exams, or tackling everyday calculations, mastering these functions on a scientific calculator is an invaluable skill.

Using Memory Functions

Using memory functions on a scientific calculator is a crucial skill that enhances your efficiency and accuracy when handling complex calculations. This guide will provide a comprehensive understanding of these functions, particularly focusing on how to leverage them effectively.

Understanding Memory Functions

Memory functions in scientific calculators allow you to store numbers or results of calculations temporarily. This feature is particularly useful when working with several values or when you need to use a specific number multiple times in different operations. The most common memory functions are:

Memory Clear (MC): Clears all data stored in the calculator's memory.
Memory Recall (MR): Retrieves the stored value from memory without clearing it.
Memory Store (MS): Saves the displayed number into memory, replacing any value that was there.
Memory Plus (M+): Adds the displayed number to the value in memory and stores the result.
Memory Minus (M-): Subtracts the displayed number from the value in memory and stores the result.

How to Use Memory Functions

Here's a step-by-step guide to using memory functions effectively on a scientific calculator:

1. **Perform Initial Calculation**: Begin by calculating the value you wish to store. For instance, if you need to add 45 and 55, enter this into your calculator and press the equals sign (=) to get the result (100).

2. **Store the Value**: To store the result (100) in memory, press the Memory Store (MS) button. Your calculator might show a small indicator, such as "M," to signify that there's a value stored in memory.

3. **Continue With New Calculations**: Proceed with other calculations as needed. For example, if you then need to multiply 20 by 5, you can do so without affecting the stored value.

4. **Recall Stored Value**: When you need to use the stored value, press the Memory Recall (MR) button. This will bring the stored number (100) back onto the screen, ready for use in your current operation.

5. **Utilize Memory Addition or Subtraction**: If you need to update the value stored in memory based on a new calculation, use the Memory Plus (M+) or Memory Minus (M-) functions. For example, if you wish to add 50

to the stored value, enter 50 on your calculator and press M+. Now, your stored value will be 150.

6. **Clear Memory**: Once you're done with the stored value and wish to clear memory, press the Memory Clear (MC) button. This will erase the stored data, allowing you to start fresh with new calculations.

Practical Tips

Multiple Memory Slots: Some advanced calculators offer multiple memory slots (labeled as M1, M2, etc.). These can be exceptionally useful for storing several different values at once. Consult your calculator's manual to understand how to use these additional slots.

Combining Memory Functions: You can combine different memory functions to manage complex calculations more efficiently. For instance, storing interim results with MS, updating them with M+ or M-, and recalling them when needed with MR.

Battery Life and Memory: Be aware that if your calculator runs on batteries and they die or are removed, you may lose the stored memory. Some models have a backup memory feature to prevent this loss.

Practice Regularly: Familiarity with these functions comes from regular use. Practice with real-life calculation scenarios to become more proficient.

By mastering the use of memory functions on your scientific calculator, you significantly enhance your ability to tackle complex mathematical problems efficiently. Whether you're a student, a professional, or just someone who loves math, these skills are invaluable.

Working with Percentages and Fractions

Working with percentages and fractions is a fundamental skill in mathematics, finance, and various other fields. Understanding how to use a scientific calculator to perform operations involving these numerical formats can greatly enhance efficiency and accuracy. Here's a comprehensive guide on the basics of handling percentages and fractions with a scientific calculator.

Understanding Percentages

1. **Conversion to and from Percentages**:
- To convert a decimal to a percentage on a scientific calculator, multiply the decimal by 100. For instance, to convert 0.75 to a percentage, input `0.75  100 = 75%`.
- To convert a percentage to a decimal, divide by 100. For example, to change 50% to a decimal, input `50 ÷ 100 = 0.5`.

2. **Percentage Increase/Decrease**:
- To calculate a percentage increase, use the formula: `((New Value - Original Value) ÷ Original Value)  100`.
- For a decrease, the same formula applies but will yield a negative result if the new value is less than the original value.

3. **Using Percentages in Calculations**:
- To add a percentage to a number, calculate the percentage of the number and then add it to the original number. E.g., to increase 200 by 15%, first calculate `200 15% = 30`, then add it back to 200.

Working with Fractions

1. **Inputting Fractions**:
- Scientific calculators often have a dedicated button for entering fractions, typically labeled as `a b/c` or `□/□`. Use this button to input fractions directly.
- Some models require you to input fractions by entering the numerator, pressing the fraction button, and then entering the denominator.

2. **Conversion Between Mixed Numbers and Improper Fractions**:
- To convert a mixed number to an improper fraction, multiply the whole number by the denominator, add the numerator, and place this result over the original denominator.
- Converting back, divide the numerator by the denominator for the whole number, and use the remainder as the new numerator.

3. **Simplifying Fractions**:
- Many calculators have a `simplify` or `reduce` function that automatically reduces fractions to their simplest

form. This is often found in the function (Fn) menu or similar.

4. **Operations with Fractions**:
- Addition, subtraction, multiplication, and division of fractions can be directly inputted into the calculator using the fraction key. The calculator will display results either as fractions or decimals, depending on its settings and the operation.

Calculator Settings and Tips

Mode Setting: Ensure your calculator is set to the correct mode for handling fractions and decimals. This can typically be found in the mode settings.
Decimal to Fraction Conversion: Some calculators allow you to convert decimals to fractions by pressing a specific button, often labeled `d/c` or `Frac`.
Percent Button: Utilize the percent (%) button for quick calculations involving percentages. This button automatically divides the entered number by 100, simplifying the conversion process.

Practice Makes Perfect

The best way to become proficient in using a scientific calculator for percentages and fractions is through practice. Try solving various types of problems to familiarize yourself with the calculator's functions and layout. Remember, the specific functions and button labels may vary between calculator models, so consult your calculator's manual for detailed instructions related to your device.

Understanding these basics will empower you to handle a wide range of mathematical problems more efficiently, making a scientific calculator an invaluable tool in your arsenal.

Introduction to Scientific Notation

Scientific notation is a method of expressing numbers that are too large or too small to be conveniently written in decimal form. It's especially useful in sciences and engineering to simplify calculations and to express numbers in a more manageable form. Understanding how to use scientific notation is essential when working with scientific calculators, as it allows for efficient and accurate computation of complex equations. This guide introduces the basics of scientific notation and its application in using a scientific calculator.

What is Scientific Notation?

Scientific notation expresses numbers as a product of two numbers: a coefficient and the power of 10. The coefficient is a number greater than or equal to 1 and less than 10, and the power of 10 is an exponent that shows how many times the coefficient should be multiplied by 10. The general form is $a \times 10^b$, where a is the coefficient and b is the exponent.

Why Use Scientific Notation?

1. **Simplifies Large and Small Numbers**: It makes it easier to read, write, and understand very large or very small numbers.
2. **Standardizes Measurements**: Scientific notation provides a uniform way of expressing measurements in scientific works.
3. **Improves Calculation Accuracy**: Reduces the risk of error when multiplying or dividing large numbers.
4. **Efficient Data Entry on Calculators**: Scientific calculators often require data to be entered in scientific notation for complex calculations.

How to Convert to Scientific Notation

To convert a number to scientific notation:
1. **Move the decimal point** in the number until it is just to the right of the first non-zero digit.
2. **Count the number of places** you moved the decimal point. This number will be the exponent b on the 10. If you moved the decimal to the left, b is positive; if to the right, b is negative.
3. **Write the number** as a product of the new coefficient and 10 raised to the power of b.

Using Scientific Notation on a Scientific Calculator

1. **Entering Numbers**: To enter a number in scientific notation on a scientific calculator, you typically use a button labeled "EXP" or "EE" for entering the exponent. For example, to input 3.2×10^4, you would press `3.2`, then `EXP` or `EE`, followed by `4`.
2. **Calculations**: Once numbers are entered in scientific notation, you can perform calculations (add, subtract, multiply, divide) as usual. The calculator automatically handles the powers of 10.
3. **Reading Results**: The calculator may display results in scientific notation automatically, especially for very large or very small numbers. Understanding scientific notation is key to interpreting these results correctly.
4. **Adjusting Display**: Some calculators allow you to switch between standard decimal notation and scientific notation for the display of results. This can usually be done through the settings or mode functions.

Practice and Applications

Using scientific notation on a calculator efficiently requires practice. Try converting numbers into scientific notation manually, then input them into your calculator to perform operations. This will help you become familiar with your calculator's specific keys and functions related to scientific notation.

In summary, mastering scientific notation and its use on scientific calculators enhances precision and efficiency in scientific calculations. Whether you're a student, educator, or professional, becoming comfortable with this notation will greatly benefit your work in the sciences and engineering.

Chapter: 3 Advanced Mathematical Functions

Trigonometric Functions: Sine, Cosine, Tangent, and Their Inverses

Understanding trigonometric functions and their inverses is a cornerstone of advanced mathematics, particularly in fields such as engineering, physics, and geometry. Trigonometric functions include sine (sin), cosine (cos), and tangent (tan), each of which relates the angles of a triangle to the lengths of its sides. Their inverses—arcsine ($\sin^{-1}$), arccosine ($\cos^{-1}$), and arctangent ($\tan^{-1}$)—allow us to work backward from the ratios of sides to the angles. Knowing how to use these functions on a scientific calculator is essential for efficiently solving problems involving trigonometry.

1. Sine, Cosine, and Tangent

Sine (sin)
The sine of an angle in a right triangle is the ratio of the length of the opposite side to the length of the hypotenuse. On your scientific calculator, you can calculate the sine of an angle by entering the angle (in degrees or radians, depending on your calculator's settings) and pressing the sin button.

Cosine (cos)
Cosine relates the length of the adjacent side to the hypotenuse. To find the cosine of an angle, enter the angle on your scientific calculator and press the cos button.

Tangent (tan)
Tangent is the ratio of the opposite side to the adjacent side. To calculate the tangent of an angle, enter the angle into your calculator and press the tan button.

2. Inverse Trigonometric Functions

Inverse trigonometric functions allow you to find an angle when you know the sides' ratios.

Arcsine ($\sin^{-1}$)
Use arcsine to find an angle when you know the opposite side and hypotenuse. Enter the ratio into the calculator and press the $\sin^{-1}$ or asin button.

Arccosine ($\cos^{-1}$)
Arccosine helps you find an angle when you have the adjacent side and hypotenuse. Enter the ratio and press the $\cos^{-1}$ or acos button on the calculator.

Arctangent ($\tan^{-1}$)

Use arctangent when you know the opposite and adjacent sides. Enter the ratio and press the $\tan^{-1}$ or atan button.

Using a Scientific Calculator

1. **Mode Selection**: Ensure your calculator is in the correct mode for the problem at hand (degree vs. radian mode). This often involves toggling a mode setting or accessing a setup menu.

2. **Entering Values**: When using trigonometric functions, input the angle directly followed by pressing the appropriate trig function key. For inverse functions, input the ratio of the sides and then press the corresponding inverse function key.

3. **Angle Units**: Pay attention to the angle units your calculator uses (degrees, radians, or grads). You might need to convert between these units, which can typically be done using specific functions on the calculator.

4. **Error Messages**: If you enter a value that doesn't make sense (like trying to find the arcsine of a number greater than 1), your calculator might display an error message. Double-check your inputs in such cases.

5. **Practice**: Familiarity with your scientific calculator's functions and layout is crucial. Practice using these functions in different contexts to gain fluency.

6. **Advanced Feature**s: Some scientific calculators also offer features to solve equations directly or perform calculations with variables. Explore these features to enhance your problem-solving capabilities.

Understanding how to effectively use these trigonometric functions and their inverses on a scientific calculator can dramatically improve problem-solving efficiency in many areas of mathematics and science. Practice regularly to become proficient and confident in their application.

Logarithmic and Exponential Functions

Logarithmic and exponential functions are two of the most important mathematical concepts, especially in fields like engineering, physics, and finance. A scientific calculator, which is a must-have tool for students and professionals alike, offers functionality to easily compute these functions. Understanding how to use these features can significantly enhance problem-solving efficiency.

Understanding Logarithmic Functions

A logarithm function is the inverse of an exponential function. It answers the question: to what exponent should the base be raised to produce a given number? The logarithmic function is typically written as $\log_b(x) = y$, which means that $b^y = x$. In scientific calculators, two types of logarithms are commonly used: the natural logarithm ($\ln$), which has a base e (where e is approximately 2.71828), and the common logarithm ($\log$), with a base of 10.

Using a Scientific Calculator for Logarithms:

1. **Common Logarithm ($\log$)**: To find the common logarithm of a number, you typically press the "log" button followed by the number. For example, to find $\log(100)$, press "log" then "100", which will display 2, because $10^2 = 100$.

2. **Natural Logarithm ($\ln$)**: To compute the natural logarithm, press the "ln" button followed by the number. For calculating $\ln(e^2)$, press "ln" then enter "e^2" (where "e" might be a separate button or require a combination like "Shift" + a number key), displaying 2, since $e^2 = e^2$.

Understanding Exponential Functions

Exponential functions involve raising a number (the base) to a power. In its simplest form, an exponential function is written as $y = b^x$, indicating that y changes exponentially as x varies. Exponential functions are crucial for modeling growth processes, such as population growth or interest compounding.

Using a Scientific Calculator for Exponential Functions:

1. **General Exponentiation**: To raise a number to a power, use the "^" or a similar button. For example, to

compute 2^3, you would enter "2", then "^", and "3", which results in 8.

2. **Natural Exponential Function (e^x)**: Most scientific calculators have a dedicated button for e, often labeled as "EXP" or "e^x". To calculate e^2, you would press this button followed by "2", yielding an approximate value of 7.389.

Tips for Effective Use

- **Understand the Order of Operations**: Scientific calculators follow the standard mathematical order of operations. Ensure you're familiar with this to avoid errors in complex calculations.

- **Utilize the Memory Functions**: For lengthy calculations involving exponential and logarithmic functions, use the calculator's memory functions (often labeled "M+", "M-", "MR", and "MC") to store interim results.

- **Practice with Real Problems**: The best way to become proficient is to use these functions while working on real-world problems. Try solving different equations or scenarios to see how these functions can be applied.

- **Check for Special Features**: Some scientific calculators have advanced features like solving for the unknown in exponential and logarithmic equations directly. Explore your calculator's manual to make full use of its capabilities.

In summary, logarithmic and exponential functions form the backbone of many scientific and mathematical applications. Mastery in using these functions on a scientific calculator can greatly enhance problem-solving skills and efficiency. Whether you're calculating the decay of a radioactive element or the future value of an investment, these functions, combined with a scientific calculator, are indispensable tools in your mathematical toolkit.

Calculating Derivatives and Integrals

Calculating derivatives and integrals is a foundational aspect of calculus, a branch of mathematics focused on rates of change and accumulation. Understanding how to compute these functions not only deepens mathematical knowledge but also enhances problem-solving skills in physics, engineering, and economics. While manual computation offers a thorough understanding, scientific calculators simplify and expedite the process. Here's a comprehensive guide on leveraging a scientific calculator for these purposes:

Understanding the Functions

1. **Derivatives**: A derivative represents the rate at which a function is changing at any given point. It is fundamental in determining slopes of tangents, velocities, and acceleration among other rates of change.

2. **Integrals**: Integrals are essentially the opposite of derivatives. They measure the total accumulation of quantities, such as areas under curves, distances, or volumes.

Using a Scientific Calculator

Calculating Derivatives

1. **Preparation**: Ensure your calculator is in the correct mode (usually found in settings), which might differ between calculators. Look for "Math" or "CAS" (Computer Algebra System) modes.

2. **Function Input**: Enter the function whose derivative you wish to calculate. This usually involves selecting the derivative function (often denoted as \(d/dx\) or similar) followed by the function expression and, lastly, the variable (commonly \(x\)).

3. **Point of Evaluation**: If you're calculating a derivative at a specific point, input that value following the function. This step computes the slope of the tangent line at that particular point on the graph.

4. **Computation**: After inputting the function and the point of interest, press the calculate or enter button. The calculator will display the derivative or the slope of the tangent at the given point.

Calculating Integrals

1. **Mode Selection**: Similar to derivatives, ensure the calculator is in the correct mode for calculating integrals.

2. **Function Input**: Input the function you wish to integrate. Access the integral function, often denoted by an elongated "S" symbol ($\int$), then enter the function followed by the differential variable (usually dx).

3. **Limits of Integration**: For definite integrals, input the lower and upper limits of integration after the function. These limits specify the range over which you're accumulating the quantity.

4. **Computation**: Hit the calculate or enter button to compute the integral. The calculator will return the area under the curve between the specified limits for definite integrals or the antiderivative function for indefinite integrals.

Tips for Effective Use

- **Familiarize with Your Calculator**: Functions and modes can vary significantly across different models and brands. Spend time understanding your specific calculator's capabilities and limitations.

- **Manual Backup**: While calculators are powerful, manually working through problems helps reinforce understanding and ensures you can tackle questions that might fall outside the calculator's capabilities.

- **Error Checking**: Always review your input for errors. A misplaced sign or parentheses can drastically alter results.

- **Consult the Manual**: For complex functions or errors, refer to your calculator's manual. Many calculators also offer online resources or forums for additional support.

In conclusion, scientific calculators are invaluable tools for computing derivatives and integrals, offering speed and precision. However, a deep understanding of the underlying mathematics is crucial for effective use and interpretation of results. As technology advances, familiarizing oneself with the latest computational tools alongside traditional methods remains a fundamental aspect of learning and applying advanced mathematics.

Complex Numbers: Basic Operations and Polar Form

Complex numbers, a fundamental concept in mathematics and engineering, extend the real numbers to include solutions to equations that have no real solution. The basic operations involving complex numbers—addition, subtraction, multiplication, and division—are pivotal in various mathematical and practical applications. The polar form of a complex number offers a different perspective, emphasizing the geometric interpretation of complex numbers. Understanding how to perform these operations using a scientific calculator can greatly enhance one's efficiency in dealing with complex numbers.

Basic Operations with Complex Numbers

Addition and Subtraction

To add or subtract complex numbers, combine the real parts and the imaginary parts separately. On a scientific calculator:
1. Enter the real part of the first complex number.
2. Use the `+` (add) or `-` (subtract) operation.
3. Enter the real part of the second complex number.

4. For the imaginary parts, repeat the steps using the imaginary unit `i` or `j` (depending on your calculator).

Multiplication
Multiplying complex numbers involves the distributive property, akin to "FOIL" in algebra. Scientific calculators typically offer a direct way to input complex numbers, usually by entering the real part, pressing a button labeled `i` or `j` for the imaginary part, and then using the multiplication operation.

Division
Division is more intricate, involving the conjugate of the denominator to eliminate the imaginary part from the denominator. Many scientific calculators automate this process:
1. Enter the numerator as a complex number.
2. Press the division operation.
3. Enter the denominator as a complex number.
4. The calculator automatically applies the conjugate method.

Polar Form and Conversion

The polar form of a complex number is an alternative representation, using a radius (magnitude) and an angle (argument). It is especially useful in trigonometry, calculus, and electrical engineering.

Conversion to Polar Form

To convert a complex number from its standard form (a + bi) to polar form (r(cosθ + isinθ)), use the formula `r = sqrt(a^2 + b^2)` for the magnitude and `θ = tan^(-1)(b/a)` for the angle. On a scientific calculator:
1. Calculate the magnitude `r` using the square root and power functions.
2. Calculate the angle `θ` using the inverse tangent function, often labeled as `tan^(-1)` or `atan`.

Operations in Polar Form

Operations like multiplication and division become simpler in polar form, reducing to multiplication/division of magnitudes and addition/subtraction of angles.

- **Multiplication**: Multiply the magnitudes and add the angles.
- **Division**: Divide the magnitudes and subtract the angles.

To perform these operations on a scientific calculator, switch to polar mode if available, and enter the magnitudes and angles as instructed. Otherwise, perform the operations on the magnitudes and angles separately using the calculator's basic functions.

Practical Tips for Using a Scientific Calculator

- **Know Your Calculator**: Familiarize yourself with the specific functions and modes of your calculator, as capabilities can vary widely.
- **Use Memory Functions**: Store intermediate results in the calculator's memory to streamline complex calculations.
- **Check the Settings**: Ensure your calculator is set to the correct mode (degree or radian) for angle measurements, especially when working with polar forms.

In conclusion, mastering basic operations and the conversion between standard and polar forms of complex numbers using a scientific calculator is invaluable for students and professionals alike. It not only saves time but also reduces the potential for errors in manual calculations, allowing for a deeper focus on analysis and interpretation of results.

Chapter: 4 Programming and Custom Functions

Introduction to Calculator Programming

"Introduction to Calculator Programming: Programming and Custom Functions in a Scientific Calculator"

Navigating the intricacies of a scientific calculator can transform it from a simple computational device to a powerful tool for solving complex problems. This transition is greatly facilitated by understanding how to program and create custom functions within your calculator. This guide offers a comprehensive introduction to calculator programming, tailored for users of scientific calculators.

Understanding Your Scientific Calculator

Before diving into programming, familiarize yourself with your calculator's basic features and capabilities. Most scientific calculators offer a range of functions beyond simple arithmetic, including statistical analysis, trigonometry, and the ability to work with variables and complex numbers. Identifying these features is the first step toward leveraging your calculator's full potential.

The Basics of Calculator Programming

Calculator programming involves creating sequences of instructions that the calculator can execute to perform specific tasks. This can range from simple macros (a series of predefined operations) to more complex programs that include conditional logic and looping.

1. **Accessing the Programming Mode**: Most scientific calculators have a specific mode for programming. This is usually accessed by pressing a designated "Program" or "PRGM" button.

2. **Writing Your First Program**: Programming on a calculator typically involves using a form of the calculator's own command language. This might involve selecting operations and functions from menus or typing out commands using a syntax specific to the calculator model.

3. **Saving and Running Programs**: After writing your program, you'll need to save it and possibly assign it to a specific button or function key. Execution is then as

simple as selecting your program and providing any necessary inputs.

Creating Custom Functions

Custom functions allow you to extend your calculator's capabilities by defining new operations that can be reused in calculations. For example, you might create a custom function to solve a particular type of equation or to perform a specific statistical analysis.

1. **Defining a Function**: This typically involves specifying a name for your function, defining any inputs (parameters) it requires, and then programming the sequence of operations that constitutes the function.

2. **Using Your Custom Functions**: Once defined, your custom function can be called just like any built-in function, passing in any required inputs. This can greatly speed up complex calculations, as you can reuse your custom function whenever needed.

Tips for Effective Calculator Programming

- **Keep It Simple**: Start with simple programs and functions to get a feel for the programming interface and syntax of your calculator.
- **Document Your Work**: Keep notes on what your programs and functions do, especially if you plan to use them in the future. It's easy to forget the specifics over time.
- **Practice Error Handling**: Learn how to debug your programs. Understanding common errors and how to fix them is crucial for creating reliable programs.
- **Explore Online Resources**: Many scientific calculators have dedicated user communities where you can find pre-written programs, tutorials, and advice on calculator programming.

Conclusion

Calculator programming can significantly enhance the functionality of your scientific calculator, turning it into a tailored tool that fits your specific needs. Whether you're solving complex equations, analyzing data, or automating repetitive calculations, understanding how to program and create custom functions is invaluable. Start simple, practice regularly, and soon you'll be unlocking the full potential of your scientific calculator.

Creating and Storing Custom Functions

Creating and storing custom functions in a scientific calculator can significantly enhance your efficiency and accuracy in solving complex problems. This guide will help you understand how to create and store these functions for easy retrieval and use.

Understanding Custom Functions

Custom functions are user-defined operations that can perform a series of calculations using different variables and constants. They are particularly useful in repetitive tasks, allowing for quicker computations without manually entering the same formula multiple times.

Why Create Custom Functions?

- **Efficiency**: Save time by automating repetitive calculations.
- **Accuracy**: Reduce the likelihood of errors in manual entry.
- **Complexity Management**: Simplify complex operations into single-step calculations.

Steps to Create and Store Custom Functions

1. Identify the Need

Before programming a custom function, clearly define the problem you're trying to solve. Determine the variables, constants, and the mathematical operations needed.

2. Access Programming Mode

Most scientific calculators have a programming mode where you can input and edit custom functions. Access this mode usually by selecting a specific function key or through the settings menu.

3. Code the Function

- **Define Variables**: Assign letters or symbols to represent variables in your formula.
- **Input Operations**: Use the calculator's keypad to enter mathematical operations and functions.
- **Use Conditional Statements**: Some calculators allow for conditional operations (if-then-else) for more complex functions.

4. Save the Function

After coding the function, save it with a unique identifier (name or number) for easy access. The saving process varies; consult your calculator's manual for specific instructions.

5. Test the Function

Always test your new function with known values to ensure it works correctly before relying on it for important calculations.

Retrieving and Using Stored Functions

To use a stored function, you typically access a library or list of custom functions through a menu in the calculator. Select your desired function, enter the required variables when prompted, and execute the operation.

Tips for Efficient Custom Functions

- **Simplicity**: Keep functions as simple as possible for ease of use and understanding.

- **Documentation**: Maintain a record of your custom functions, including their purpose and instructions for use.
- **Regular Updates**: Revise and update your functions as needed to ensure they remain accurate and efficient.

Conclusion

Creating and storing custom functions in a scientific calculator is a powerful way to enhance your problem-solving capabilities. It allows for more efficient, accurate, and simplified computations, particularly for complex or repetitive tasks. By following the steps outlined above, you can effectively develop custom functions tailored to your specific needs, making the most out of your scientific calculator's capabilities.

Using Pre-installed Applications

Using pre-installed applications, specifically in the realm of programming and custom functions, can significantly enhance the functionality and usability of a scientific calculator. This guide provides insights on how to leverage these features to perform complex calculations, automate tasks, and solve advanced mathematical problems.

Understanding Your Scientific Calculator

Before diving into programming and custom functions, familiarize yourself with your scientific calculator's basic operations, modes, and features. Most scientific calculators come equipped with standard functions such as trigonometric operations, logarithms, and power calculations. They also offer modes for different mathematical operations like complex numbers, statistics, and equation solving.

Accessing Pre-installed Applications

Scientific calculators often include pre-installed applications designed to solve specific types of problems or perform complex operations. To access these, look for an "Apps" or "Program" button on your calculator. Navigating through these menus will reveal applications ranging from polynomial root finders to matrix calculators and statistical analysis tools.

Programming and Custom Functions

Programming on a scientific calculator allows you to define custom functions, automate repetitive tasks, and even create simple games or utilities. The programming capability varies by model, but most scientific calculators support basic programming constructs such as loops, conditionals, and variables.

Creating Custom Functions

1. **Access the Programming Mode**: Find and select the option to enter the programming mode on your calculator. This is usually found in the apps or program menu.

2. **Define Your Function**: Use the calculator's syntax to write your function. Start by naming your function, then write the operations it should perform. Input and output commands will be essential for functions that interact with the user.

3. **Save and Exit**: Once your function is written, save it and exit the programming mode. Your function will now be available for use like any other built-in function.

Examples of Custom Functions

- Compound Interest Calculator: A program that calculates the future value of an investment based on the principal amount, interest rate, and time.
- Quadratic Equation Solver: A function that takes the coefficients of a quadratic equation as inputs and returns its roots, handling both real and complex solutions.

Tips for Effective Use

- **Leverage Documentation**: Many calculators come with detailed manuals or online resources that explain how to use pre-installed applications and program custom functions. Use these resources to understand the capabilities of your calculator fully.

- **Practice Regularly**: Familiarity with programming and custom functions grows with use. Practice by trying to automate or simplify the calculations you do regularly.

- **Join Communities**: Online forums and communities of calculator enthusiasts can be invaluable resources for learning tips, tricks, and getting help with programming your calculator.

- **Backup Your Work**: If your calculator supports it, regularly backup your programs and custom functions. This can usually be done by connecting your calculator to a computer.

Using pre-installed applications and programming custom functions on your scientific calculator can unlock its full potential, making it an even more powerful tool for your studies or professional work. Whether you're solving complex equations, analyzing data, or just exploring mathematics, these features can provide you with the flexibility and efficiency to achieve your goals.

Chapter: 5 Statistical Functions

Entering and Editing Data Sets

Entering and editing data sets in a scientific calculator is essential for performing statistical analysis and calculations. The process allows users to input and modify groups of numbers to calculate statistical functions such as mean, median, standard deviation, and more. This guide will walk you through the steps for entering and editing data sets using a scientific calculator, focusing on common functions and practices.

Step 1: Entering Data Mode

1. **Switch to Statistical Mode**: Most scientific calculators have a specific mode for statistical calculations. This mode is usually accessed by pressing a button labeled "STAT" or something similar. Once in this mode, you can start entering your data set.

2. **Inputting Data**: To enter data, type in a number using the keypad and then press the "Data" key (often labeled

"DAT" or "Enter") to save the number in the calculator's memory. Repeat this step for each data point in your set.

Step 2: Editing Data

1. **Reviewing Entered Data**: Some calculators allow you to scroll through the data you have entered. This feature is useful for checking your entries for mistakes. Use the arrow keys or a specific "Review" button to navigate through your data set.

2. **Editing Specific Entries**: If you find an error in your data or need to update a value, use the calculator's edit function. This function is usually accessed by navigating to the incorrect entry and then pressing an "Edit" button. After making your changes, press the "Data" key to save the updated entry.

3. **Deleting Entries**: To remove an incorrect or unnecessary data point, navigate to the entry and press a "Delete" or "Clear" button. This action removes the data point from your set.

Step 3: Calculating Statistical Functions

1. **Selecting a Statistical Function**: After entering and editing your data set, you can calculate various

statistical measures. Access the function you need (e.g., mean, standard deviation) by pressing the corresponding button on your calculator. The labeling and availability of these functions can vary by calculator model.

2. **Viewing Results**: After selecting the function, the calculator will display the result. For some calculations, you may need to press an "Equals" or "Calculate" button to execute the function.

3. **Using Advanced Statistical Functions**: For more complex statistical analyses (like regression analysis), consult your calculator's manual for specific instructions. These functions may require entering data in pairs or using additional modes/settings.

Tips for Efficient Data Management

- **Double-Check Your Entries**: Always review your data set for accuracy before performing calculations. Errors in data entry can lead to incorrect results.
- **Utilize Memory Functions**: If your calculator has memory capabilities, use them to store and recall statistical results for later use.
- **Refer to the Manual**: Scientific calculators can vary significantly in functionality and interface. For the best results, refer to your calculator's manual for detailed instructions tailored to your specific model.

Conclusion

Mastering the process of entering and editing data sets in a scientific calculator enhances your ability to perform a wide range of statistical analyses efficiently. Whether you're a student, teacher, or professional, becoming proficient with your calculator's statistical functions can save you time and improve the accuracy of your calculations. Remember, familiarity with your specific calculator's features and functions is key to leveraging its full potential in statistical analysis.

Calculating Mean, Median, Mode, and Standard Deviation

Calculating mean, median, mode, and standard deviation are fundamental statistical functions that provide insights into the distribution and variability of a dataset. When working with a scientific calculator, understanding how to perform these calculations can enhance your efficiency in data analysis. Below is a comprehensive guide on how to calculate these statistical measures using a scientific calculator:

1. Mean (Arithmetic Mean)

The mean is the average of a dataset, calculated by summing all the values and dividing by the count of the values.

How to Calculate on a Scientific Calculator:

- Step 1: Enter the first number of your dataset.
- Step 2: Press the faddition (+) key.
- Step 3: Enter the next number in your dataset.
- Step 4: Repeat Steps 2 and 3 until all numbers are entered.
- Step 5: Press the division (/) key.
- Step 6: Enter the total number of data points.
- Step 7: Press the equals (=) key to get the mean.

2. Median

The median is the middle value in a dataset when the numbers are arranged in ascending or descending order. If there's an even number of observations, the median is the average of the two middle numbers.

How to Calculate on a Scientific Calculator:
Calculating the median directly on a scientific calculator isn't straightforward as it doesn't sort or automatically select the middle value. You'll need to:
- Step 1: Manually sort your dataset in ascending or descending order.
- Step 2: Identify the middle number (or the average of the two middle numbers for an even set of values).

3. Mode

The mode is the value that appears most frequently in a dataset. There can be more than one mode in a dataset.

How to Calculate on a Scientific Calculator:
Similar to the median, there's no direct function for calculating the mode on most scientific calculators. You need to:
- Step 1: List your data set.
- Step 2: Count the frequency of each value.
- Step 3: Identify the value(s) that appear(s) most frequently.

4. Standard Deviation

Standard deviation measures the amount of variation or dispersion in a set of values.

How to Calculate on a Scientific Calculator:
- Step 1: Calculate the mean of your dataset.
- Step 2: Subtract the mean from each data point and square the result for each.
- Step 3: Sum all the squared results.
- Step 4: Divide this total by the number of data points (for a population) or by one less than the number of data points (for a sample).
- Step 5: Take the square root of the result from Step 4. This gives you the standard deviation.

Note: Some advanced scientific calculators have built-in functions for statistical calculations, allowing you to enter data points and directly compute these statistics. Consult your calculator's manual for specific instructions, as the procedure can vary significantly across different models and brands.

Conclusion

Understanding how to calculate mean, median, mode, and standard deviation is crucial for statistical analysis. While not all calculations are straightforward on a scientific calculator, knowing the basic steps for each can significantly aid in manual computations. For more complex datasets or when precision is paramount, consider using statistical software or a calculator with built-in statistical functions.

Regression Analysis and Plotting Graphs

Regression analysis is a powerful statistical method that helps in understanding the relationship between dependent (target) and independent (predictor) variables. It's widely used in various fields such as economics, engineering, and the social sciences to predict outcomes and assess the strength of relationships. When working with regression analysis, plotting graphs is crucial for visualizing these relationships and interpreting the results more effectively. Even though advanced statistical software is often used for complex analyses, understanding how to perform basic regression analysis and plotting using a scientific calculator can be invaluable for quick calculations and in situations where software is not available.

Steps to Perform Regression Analysis Using a Scientific Calculator

1. **Understand Your Data**: Identify your dependent and independent variables. The dependent variable is what you're trying to predict or explain, and the independent variable is what you think will influence the dependent variable.

2. **Choose the Right Regression Model**: For most scientific calculators, linear regression (y = mx + b) is the default and possibly only option. This model assumes a straight-line relationship between the dependent and independent variables.

3. **Input Your Data**: Enter your data points into the calculator. This usually involves going into a specific statistics mode on your calculator and entering values for your independent (x) and dependent (y) variables. The exact method will depend on your calculator model, so refer to its manual.

4. **Perform the Regression**: After entering your data, instruct the calculator to perform a regression analysis. This typically involves selecting a regression function from the calculator's menu. The calculator will then compute the regression coefficients (slope m and y-intercept b in the case of linear regression), which define the best-fit line through your data.

5. **Interpret the Results**: The calculator will provide you with the regression coefficients and possibly other statistics like the correlation coefficient (R) or the coefficient of determination (R^2). The slope (m) indicates the direction and steepness of the line, while the intercept (b) shows where the line crosses the y-axis. R^2 tells you how much of the variation in your dependent variable is explained by your model.

Plotting Graphs on a Scientific Calculator

Plotting graphs directly on a scientific calculator is generally limited to more advanced models with graphing capabilities. If your calculator has these capabilities, you can use the regression equation obtained from your analysis to plot the line of best fit:

1. **Switch to Graphing Mode**: Access the graphing function on your calculator.
2. **Enter the Regression Equation**: Input the regression equation you derived from your analysis into the equation editor of the graphing function. Use the calculated slope and intercept values.
3. **Adjust the Viewing Window**: Set appropriate limits for the x and y axes to ensure that your data points and the regression line are visible on the graph.
4. **Plot and Analyze**: Execute the command to plot the graph. You should see your data points and the regression line. This visual representation helps in understanding how well the line fits the data.

Tips for Using a Scientific Calculator for Regression Analysis

- Familiarize Yourself with Your Calculator: Different models have different ways of entering data and

performing calculations. Spend time learning these specifics.
- **Check for Accuracy**: Always double-check your data entry to avoid errors in your analysis.
- **Use Software for Complex Analyses**: While scientific calculators are useful for on-the-go analysis and educational purposes, use statistical software for more complex analyses involving multiple variables or non-linear relationships.

In summary, while scientific calculators are somewhat limited in their capabilities for performing advanced statistical analyses, they can be incredibly useful tools for conducting basic regression analysis and plotting. Understanding how to leverage these functions can enhance your analytical skills and provide quick insights into data relationships, even without access to specialized software.

Chapter: 6 Graphing Capabilities

Understanding the Graphing Interface

Understanding the Graphing Interface in relation to using a scientific calculator involves navigating the features and functionalities that allow users to visually represent mathematical equations and data. This capability is instrumental in a wide range of academic and professional fields, including mathematics, engineering, and the sciences, providing a visual insight into complex problems. Here's a detailed guide on how to use these graphing capabilities effectively.

Familiarize with Your Calculator's Capabilities

First, identify whether your scientific calculator has graphing capabilities. Advanced scientific calculators often come equipped with a graphing feature, but the extent of these capabilities can vary. Models like the TI-84 Plus or Casio fx-9860GII are examples of calculators with robust graphing functionalities.

Accessing the Graphing Mode

To begin graphing, switch your calculator to its graphing mode. This usually involves pressing a specific key or combination of keys. For instance, on a TI-84 Plus, you might press the "Graph" button to enter the graphing mode.

Entering Equations

Once in graphing mode, you will need to enter the equation you wish to graph. This is typically done through an equation editor, where you can input mathematical expressions. Make sure to familiarize yourself with how your calculator handles different functions and variables.

Setting the View Window

Before graphing, adjust the view window to ensure that the relevant parts of the graph are visible. This involves setting the minimum and maximum values for both the x-axis and y-axis. Some calculators offer automatic adjustments based on the entered equation, but manual adjustment can provide more precise control.

Graphing the Equation

After setting up your equation and view window, initiate the graphing process. Your calculator will plot the equation based on the settings you've provided. For complex equations, this might take a few moments.

Analyzing the Graph

Most graphing calculators offer tools to analyze the graphed equation. These can include finding the roots (zeroes), maxima and minima, and intersections with other graphs. Access these features typically through a menu in the graphing mode.

Zoom and Trace Functions

Use zoom functions to get a closer look at specific parts of your graph, and the trace function to move along the graph and obtain coordinates of specific points. These features are invaluable for detailed analysis of the graphed data.

Saving and Recalling Graphs

Some calculators allow you to save and recall graphs for future reference. This is especially useful for comparing multiple graphs or for later analysis.

Practice and Experimentation

The best way to become proficient with the graphing interface on your scientific calculator is through practice. Experiment with different types of equations to see how they are represented graphically. This hands-on experience will improve your understanding of both the mathematical concepts and the calculator's capabilities.

Troubleshooting

If you encounter issues or unexpected results while using the graphing features, consult your calculator's manual. Most problems can be resolved by adjusting settings or understanding more about how your calculator interprets input.

Conclusion

Graphing capabilities on scientific calculators are a powerful tool for visualizing and analyzing mathematical equations. By becoming familiar with your calculator's specific features and functions, you can leverage this tool to enhance your understanding of complex mathematical concepts. Whether you are a student, educator, or professional, mastering the graphing interface of your scientific calculator can significantly contribute to your mathematical toolkit.

Plotting and Analyzing Functions

Plotting and analyzing functions are fundamental capabilities in mathematics, physics, engineering, and numerous other scientific disciplines. Understanding how to use a scientific calculator to perform these tasks can significantly enhance one's efficiency and accuracy in solving complex problems. This comprehensive guide provides an overview of how to use the graphing capabilities of a scientific calculator to plot and analyze functions.

Understanding Functions

A function is a relationship between a set of inputs and a set of possible outputs where each input is related to exactly one output. Functions are often expressed in the form of f(x), where x represents the input value.

Scientific Calculators with Graphing Capabilities

Not all scientific calculators have graphing capabilities. Graphing calculators are a subset of scientific calculators that offer the ability to plot graphs of functions, equations, and inequalities. Examples of popular graphing calculators include the TI-84 Plus, Casio fx-9860GII, and HP Prime.

Basic Steps to Plot a Function

1. **Enter the Function**: Access the graphing mode on your calculator. This usually involves pressing a dedicated "Graph" button or navigating through a menu. Once in graphing mode, enter the function you wish to plot. This typically involves typing the equation exactly as it appears, using the calculator's syntax for mathematical operations.

2. **Adjust the Viewing Window**: Before plotting, it's crucial to adjust the viewing window to ensure that the significant parts of the function are visible. This might involve setting the minimum and maximum values for both the x-axis (horizontal) and the y-axis (vertical). Adjust these settings based on the expected behavior of the function.

3. **Plot the Graph**: With the function entered and the viewing window set, plot the graph by pressing the "Graph" or "Draw" button. The calculator will display the graph on its screen.

Analyzing Functions

After plotting the function, graphing calculators offer various tools to analyze the graph:

- **Find Roots/Zeros**: These are the points where the function crosses the x-axis. Many calculators have a "Zero" function that allows you to approximate these points.

- **Determine the Vertex**: For quadratic functions, the vertex represents the highest or lowest point on the graph. Calculators usually have a function to find this point directly.

- **Calculate Intercepts**: You can find where the function intersects the y-axis by using the calculator's built-in functions to evaluate the function at x=0.

- **Analyze Slope and Tangents**: Some calculators can calculate the slope of the tangent line to the curve at a given point, which is essential for understanding the function's behavior at that point.

- **Trace the Function**: This feature allows you to move along the curve and see the coordinates of specific points. It's useful for getting a general sense of the function's shape and behavior.

Tips for Effective Graphing and Analysis

- **Familiarize Yourself with Your Calculator**: Different calculators may have different methods for entering functions and analyzing graphs. Spend some time learning these features.

- **Use the Manual**: If you're unsure how to access or use the graphing and analysis features, consult your calculator's manual. These documents are often rich with information and examples.

- **Practice with Different Functions**: The best way to become proficient is to practice plotting and analyzing a variety of functions. Experiment with linear, quadratic, polynomial, exponential, and trigonometric functions to understand how they behave graphically.

- **Check Your Work Algebraically**: Whenever possible, verify the results obtained from your calculator algebraically or with another method. This practice helps ensure accuracy and reinforces your understanding of the concepts.

Plotting and analyzing functions using a scientific calculator with graphing capabilities is a powerful skill that can enhance your understanding of mathematical concepts and improve your problem-solving abilities. By following these guidelines and practicing regularly, you'll be able to effectively utilize your calculator to its full potential.

Using Zoom and Trace Features

Using the Zoom and Trace features on a scientific calculator, especially those with graphing capabilities, enhances the understanding and interpretation of graphical data significantly. Here's a comprehensive guide on how to utilize these functions effectively.

Understanding Graphing on Scientific Calculators

Graphing calculators allow you to plot equations and inequalities, offering a visual representation of functions and their characteristics. Before diving into the Zoom and Trace features, ensure you're familiar with entering equations and setting up the graphing window on your specific calculator model.

Zoom Function

The Zoom feature adjusts the viewing window of your graph, enabling you to examine different parts of the graph in greater detail or to view the graph more broadly to understand its general shape and intercepts.

How to Use the Zoom Feature:

1. **Plot Your Function**: Enter the function you wish to graph according to your calculator's instructions.
2. **Access the Zoom Menu**: Look for a button labeled "Zoom" and press it. This will open a menu with several zooming options.
3. **Choose Your Zoom Option**: Common options include Zoom In, Zoom Out, Zoom Fit, and Zoom Square. Each serves a different purpose:

   - **Zoom In**: Allows you to see a smaller portion of the graph in more detail. Use the arrow keys to move the cursor to the area you want to zoom in on before selecting this option.
   - **Zoom Out**: Expands the view to include a larger portion of the graph. This is useful for understanding the overall shape of the graph.
   - **Zoom Fit**: Adjusts the viewing window to fit the graph optimally within the screen, showing the most relevant portions of the graph.
   - **Zoom Square**: Makes the scale of the x-axis equal to the y-axis, which can be particularly useful for graphs where proportionality is important.

Trace Function

The Trace feature allows you to move along the graph and observe the exact coordinates of specific points. This is incredibly useful for identifying key features of the graph, such as intercepts, turning points, and areas of interest.

How to Use the Trace Feature:

1. **Graph Your Function**: After plotting your function, you can start tracing it.
2. **Activate the Trace Feature**: Press the "Trace" button on your calculator.
3. **Navigate the Graph**: Use the left and right arrow keys to move along the graph. As you move, the calculator will display the coordinates of the current location on the graph.
4. **Identify Key Points**: As you trace the graph, look for important features. The exact coordinates of these points will be shown, allowing for precise analysis.

Tips for Effective Use

- **Experiment with Different Views**: Use both Zoom and Trace features in tandem to get both a macro and micro view of your graphs.
- **Documentation**: Keep a notebook handy to jot down important coordinates and observations as you explore different sections of the graph.
- **Practice with Various Functions**: Familiarize yourself with these features by practicing with a wide range of functions to understand how different types of graphs behave.

Conclusion

The Zoom and Trace functions of a scientific calculator with graphing capabilities are powerful tools for exploring and understanding mathematical functions and their graphical representations. By mastering these features, you can gain deeper insights into your mathematical investigations, making these calculators an indispensable tool in both educational and professional settings. Always refer to your calculator's manual for model-specific instructions, as the operation of these features can vary between different calculators.

Saving and Exporting Graphs

Saving and exporting graphs from a scientific calculator is a powerful feature that enhances understanding and communication of mathematical concepts. This functionality is particularly beneficial in educational and professional settings where visual representation of data is crucial. Here's a comprehensive guide on how to utilize graphing capabilities in a scientific calculator, focusing on saving and exporting graphs.

Understanding Graphing Capabilities

Most advanced scientific calculators come with graphing capabilities, allowing users to plot equations and visualize mathematical functions. These calculators can handle a variety of functions, including linear, polynomial, trigonometric, and exponential.

Steps to Graph a Function

1. **Accessing the Graphing Function**: Navigate to the graphing mode on your calculator. This usually involves selecting the 'Graph' button or accessing a menu and choosing a graphing option.
2. **Entering Equations**: Input the mathematical equation you wish to graph. Some calculators allow for

multiple equations to be entered and graphed simultaneously.

3. **Setting the Viewing Window**: Adjust the viewing window to ensure that the key features of the graph are visible. This may involve setting the x and y-axis limits and scaling.

4. **Plotting the Graph**: Once the equation is entered and the viewing window is set, plot the graph. The calculator will display the graph on the screen.

Saving Graphs

Saving a graph on a scientific calculator allows you to revisit the graph without needing to re-enter the equation or re-adjust the viewing settings.

1. **Using the Save Function**: Look for a save option in the graphing menu. The process may vary, but generally, you can select the 'Save' or 'Memory' option.

2. **Naming the Graph**: Some calculators allow you to name the saved graph for easy identification later.

3. **Memory Considerations**: Be aware of the calculator's memory limitations. Older or entry-level models might have limited storage, affecting the number of graphs that can be saved.

Exporting Graphs

Exporting graphs enables sharing with others or inserting into documents and presentations. Export capabilities depend on the calculator model and connectivity options.

1. **Connectivity Options**: Some calculators come with USB ports or Bluetooth capabilities, allowing for direct connection to computers or other devices.
2. **Exporting the Graph**: Use the calculator's software (if available) on your computer to export the graph. This might involve transferring a screenshot of the graph or using a specific export function.
3. **File Formats**: Graphs are usually exported in common image formats such as PNG or JPEG, or as PDF documents, depending on the calculator's software capabilities.

Tips for Effective Use

- **Experiment with Graphing Features**: Spend time learning the graphing functions of your calculator to make the most out of its capabilities.
- **Consult the Manual**: For specific instructions on saving and exporting graphs, consult your calculator's user manual as processes can vary significantly between models.

- **Use Graphs to Enhance Learning**: Graphs can provide visual insights into mathematical concepts, making them invaluable for learning and teaching.

Conclusion

Graphing capabilities in scientific calculators are indispensable tools for visualizing and analyzing mathematical functions. Being adept at saving and exporting these graphs can significantly enhance productivity and communication in both educational and professional contexts. By following the above steps and tips, users can effectively utilize these features to their full potential.

Chapter: 7 Troubleshooting and Maintenance

Common Issues and How to Solve Them

Using a scientific calculator can sometimes be challenging, especially for new users or those encountering complex mathematical problems. Below is a detailed guide covering common issues and their solutions, focusing on troubleshooting and maintenance for optimal use of a scientific calculator.

1. Understanding Complex Functions

Issue: Difficulty in using functions like trigonometry, logarithms, exponents, and matrix operations.
Solution: Familiarize yourself with the instruction manual of your specific model. Practice entering simple equations to understand how your calculator interprets and prioritizes operations. Many calculators follow the order of operations (PEMDAS), but checking your manual and doing practice problems can clarify many issues.

2. Battery Problems

Issue: The calculator turns off unexpectedly or fails to turn on.
Solution: Replace the batteries regularly, even if the device is not indicating low power. For calculators with a solar panel, ensure it's exposed to sufficient light. Clean the battery contacts and compartment to ensure a good connection if the calculator still doesn't turn on after battery replacement.

3. Error Messages

Issue: Receiving error messages when trying to solve problems.
Solution: Error messages typically indicate input mistakes or mathematical errors (like division by zero). Review the equation to ensure it's entered correctly. Refer to the calculator's manual to understand specific error codes and how to resolve them.

4. Display Issues

Issue: The screen is too dim, or the information is not fully displayed.
Solution: Adjust the contrast if your calculator allows it, usually through a specific button combination (check your manual). If the display is partially visible or has lines running through it, it might indicate a more serious hardware issue requiring professional repair.

5. Sticking Buttons

Issue: Buttons not registering inputs consistently.
Solution: Lightly clean around the buttons with a soft, dry brush or compressed air. Avoid using liquids that could seep into the calculator. If the issue persists, it may require professional cleaning or repair.

6. Incorrect Results

Issue: The calculator provides unexpected results for operations.
Solution: Ensure you're in the correct mode for the operation you're performing (e.g., degree vs. radian mode for trigonometric functions). Reset the calculator to its default settings as per the manual's instructions if problems persist, as this can resolve hidden settings issues.

7. Mode Confusion

Issue: Calculator is set to an incorrect mode, affecting results (e.g., statistical, scientific, or regression modes).
Solution: Review the manual to understand how to switch between modes and ensure you're in the appropriate mode for your calculations. Practice switching modes to become more familiar with the process.

8. Memory Errors

Issue: Calculator runs out of memory when performing complex operations or storing data.
Solution: Clear the calculator's memory of unnecessary data or variables. Refer to the manual on how to perform this task without losing important information.

Maintenance Tips

- **Regular Cleaning**: Keep your calculator clean by wiping it with a soft, dry cloth. Avoid using harsh chemicals.
- **Storage**: Store your calculator in a protective case and avoid exposing it to extreme temperatures or moisture.
- **Updates**: For calculators that support firmware updates, ensure they are up to date to benefit from improved functionalities and bug fixes.

In summary, most issues with scientific calculators can be resolved by referring to the instruction manual, practicing regular maintenance, and understanding the specific functions and modes of your device. For more complex problems, consider contacting the manufacturer's customer support.

Updating Calculator Firmware

Updating the firmware on your scientific calculator is a critical step in ensuring it performs optimally, unlocking new features, and fixing any bugs or issues that might affect its functionality. This guide will provide a comprehensive overview of the process, focusing on the importance of firmware updates, the steps involved, troubleshooting common issues, and maintenance tips to keep your calculator in top condition.

Importance of Firmware Updates

1. **Performance Improvements**: Firmware updates often include optimizations that make your calculator faster and more responsive.
2. **New Features**: Manufacturers sometimes add new functions and capabilities through updates, extending your calculator's utility.
3. **Bug Fixes**: Updates can resolve known glitches or issues, ensuring your calculator works as intended.
4. **Compatibility**: Keeping your firmware up to date ensures compatibility with external software tools or apps designed for scientific calculations.

How to Update Your Calculator's Firmware

1. **Check Compatibility and Requirements**: Ensure your calculator model supports firmware updates. Visit the manufacturer's website for compatibility information and specific requirements, such as needing a computer or specific cables.

2. **Backup Your Data**: Before proceeding, back up any important data stored on your calculator. Firmware updates can potentially erase stored information.

3. **Download Firmware Update**: Visit the official website of your calculator's brand to download the latest firmware version. Make sure to choose the version that corresponds to your calculator model.

4. **Connect Your Calculator to a Computer**: Use the recommended cable (usually USB) to connect your calculator to your computer. The specific steps may vary, so refer to your calculator's manual for detailed instructions.

5. **Install the Update**: Launch the firmware update tool you downloaded and follow the on-screen instructions to begin the update process. This typically involves selecting your calculator model and the downloaded firmware file, then clicking on an "update" button.

6. **Finalize the Update**: Once the update process is complete, disconnect your calculator safely. It may automatically restart, and in some cases, you might need to reset it to its factory settings.

Troubleshooting Common Issues

1. **Update Fails to Install**: Ensure your calculator is correctly connected and recognized by your computer. Check for any specific software that needs to be installed on your computer for the calculator to communicate properly.

2. **Calculator Not Recognized**: Try using a different USB port or cable. Make sure any necessary drivers are installed on your computer.

3. **Data Loss**: If you've lost data after an update, restore the backup you made. If you didn't back up, unfortunately, the data cannot be recovered. Always backup before updating.

4. **Functions Not Working Properly**: Perform a reset on your calculator, as some updates may require it to function correctly. Refer to your manual for how to do this safely.

Maintenance Tips

1. **Regular Updates**: Regularly check for firmware updates to ensure your calculator is always running the latest version.

2. **Protective Case**: Use a protective case to shield your calculator from physical damage and to keep it clean.

3. **Battery Care**: Replace batteries promptly or recharge them as necessary to avoid power issues that could interrupt the update process.

4. **Cleaning**: Clean your calculator's exterior with a soft, dry cloth. Avoid using harsh chemicals or liquids.

Updating your calculator's firmware is a straightforward process that can significantly enhance its performance and longevity. By following these steps, troubleshooting common issues, and adhering to maintenance best practices, you can ensure that your scientific calculator remains a reliable tool for all your calculation needs.

Cleaning and Caring for Your Calculator

Cleaning and caring for your scientific calculator is essential to ensure its longevity and optimal performance. Scientific calculators, with their advanced functions and capabilities, are invaluable tools for students, professionals, and anyone involved in complex mathematical, engineering, or scientific tasks. Proper maintenance and troubleshooting can prevent malfunctions and ensure the device remains reliable. Here's a comprehensive guide on how to keep your scientific calculator in top condition.

Regular Cleaning

1. **Exterior Cleaning:**

- **Frequency**: Clean the exterior of your calculator at least once a month or more often if it's used frequently.
- **Materials Needed**: Use a soft, lint-free cloth. You can slightly dampen the cloth with water or a mild cleaning solution.
- **Procedure**: Gently wipe the surface, keys, and screen. Avoid using abrasive cloths or harsh chemicals that can damage the screen or the calculator's casing.

2. **Keyboard Cleaning**:

- **For Stuck Buttons**: If some buttons are not responding properly, it could be due to dust or debris. Use a can of compressed air to gently blow away any particles from around the buttons.
- **For Sticky Buttons**: Use a cotton swab dipped in isopropyl alcohol to carefully clean around the sticky buttons. Avoid letting the liquid seep into the calculator.

3. **Screen Care**:

- **Scratch Prevention**: Use a screen protector to prevent scratches. If the screen is already scratched, avoid using substances to fill in or "repair" the scratches, as these can cause more damage.

Maintenance

1. **Battery Replacement**:

- **Signs for Replacement**: Dimming display or non-responsive calculator even after cleaning.
- **Procedure**: Replace the batteries at least once a year or as per the manufacturer's recommendation. Ensure to insert the batteries correctly according to the polarity marked inside the battery compartment. For solar-powered calculators, ensure they are exposed to sufficient light.

2. **Storage**:
- Store your calculator in a protective case to prevent dust accumulation and protect it from physical damage. Avoid extreme temperatures and direct sunlight, which can damage the electronics and screen.

Troubleshooting

1. **Calculator Not Turning On**:
- Check the batteries and replace them if necessary. For solar-powered models, try exposing them to light for a while.
- Press the reset button if available, using a paperclip or a similar tool. This can fix minor glitches.

2. **Unresponsive or Erratic Behavior**:
- Perform a soft reset as described in the user manual. This often resolves unexplained errors or issues without affecting stored data.
- If problems persist, consult the user manual for a hard reset procedure, but be aware this may erase all stored data.

3. **Display Issues**:
- If the display is too dim or too bright, adjust the contrast. This can usually be done through a simple key combination found in the user manual.

- For garbled or incomplete display, a reset is often helpful. Check for specific troubleshooting steps in the user manual related to display issues.

4. **Error Messages**:
- Consult the user manual for specific error codes and solutions. Common issues may include syntax errors in input or memory limitations.

How to Use a Scientific Calculator Efficiently

- **Familiarize Yourself with the Manual**: Understanding the specific functions, operations, and troubleshooting tips for your model is crucial.
- **Regular Practice**: Regular use and practice with your calculator will help you become more efficient and familiar with its functions.
- **Keep It Simple**: Start with basic operations and gradually move on to more complex calculations as you become more comfortable with the device.
- **Utilize Memory Functions**: Learn how to use and clear the calculator's memory functions to streamline your calculations.

In summary, proper cleaning, careful maintenance, and adept troubleshooting are key to extending the life of your scientific calculator and ensuring its reliability for all your calculation needs. Remember, the manual is your best resource for specific care instructions and troubleshooting advice tailored to your model.

Chapter: 8 Tips and Tricks

Shortcuts for Efficient Calculation

"Shortcuts for Efficient Calculation: Mastering the Scientific Calculator"

Navigating the world of mathematics and science often requires more than just understanding theories and formulas; it also demands the ability to calculate efficiently. This is where a scientific calculator becomes an indispensable tool. Here's a comprehensive guide on shortcuts for efficient calculation, tailored to help you make the most out of your scientific calculator.

Understanding Your Scientific Calculator:
Before diving into shortcuts, familiarize yourself with your scientific calculator. Although models vary, most share common functionalities. Know where the basic operations are, along with more complex functions like trigonometric operations, logarithms, exponents, and possibly programming capabilities.

1. Use Memory Functions:
Memory functions (M+, M-, MR, MC) are time-savers. Store a number you frequently use in the calculator's memory to recall it instantly for later calculations. This is

especially useful for constants in physics or repeated values in an equation.

2. **Master the Use of Parentheses**:
Correctly using parentheses can significantly speed up your calculations and ensure accuracy. By grouping parts of an equation, you can override the default order of operations, allowing you to input complex equations without breaking them down step by step.

3. **Shortcut Keys for Functions**:
Many scientific calculators have shortcut keys for commonly used functions. For instance, accessing π or e doesn't always require navigating through menus. Learning these shortcuts can drastically reduce calculation time.

4. **Utilize the Scientific Notation**:
For very large or small numbers, use your calculator's scientific notation feature (often labeled as "EXP" or "EE"). This not only speeds up input but also ensures precision in your calculations.

5. **Repeat Last Calculation**:
Many calculators allow you to repeat the last operation by pressing a certain key (often "ANS" or "="). This is useful for iterative calculations, where you're applying the same operation to a series of numbers.

6. **Make Use of Built-in Functions and Constants**:

Take advantage of built-in functions for trigonometry, logarithms, and exponentials, which are more accurate and faster than manual calculations. Similarly, using pre-stored constants (like the speed of light, c, or gravitational constant, G) can save time.

7. Explore Programming Capabilities:
Some advanced scientific calculators allow for programming of custom functions or sequences. If you find yourself repeatedly performing a specific calculation, programming it into your calculator can save immense time.

8. Familiarize with Conversion Features:
Convert between different units or number bases using your calculator's conversion functions. This is particularly handy in physics and chemistry calculations.

9. Utilize Graphing Functions for Visualization:
If your calculator has graphing capabilities, use it to visualize functions for better understanding or to find specific values (roots, maxima, minima) more efficiently than solving algebraically.

10. Practice!
Like any skill, proficiency with a scientific calculator comes from practice. Familiarize yourself with its functions and shortcuts in the context of your regular work or study to become more efficient.

Conclusion:

Mastering the use of a scientific calculator through these shortcuts and tips can dramatically enhance your efficiency in calculations, allowing you more time to focus on understanding concepts rather than getting bogged down in computations. Whether you're solving complex equations, exploring trigonometric identities, or calculating statistical data, these strategies will serve as invaluable tools in your mathematical toolkit.

Recommended Resources for Learning

Learning how to use a scientific calculator effectively can greatly enhance your understanding and efficiency in solving mathematical and scientific problems. This skill is invaluable for students, educators, and professionals in various fields. Here, we provide a comprehensive guide on resources and tips to master the use of a scientific calculator.

1. Understand the Basics

First, get familiar with your calculator's manual. Most scientific calculators come with a detailed manual explaining all the functions and features. This is your best starting point.

Resources:
- **Calculator's Manual**: The first and most important resource. If you've lost yours, manufacturers often provide digital copies on their websites.
- **Manufacturer's Website:** Companies like Casio, Texas Instruments, and HP offer tutorials and guides on their websites.

2. Online Tutorials and Courses

There are numerous free and paid courses and video tutorials available online that cater to all levels of learners.

Resources:
- **YouTube**: Channels like Khan Academy, MathAndScience, and The Calculator Guide offer step-by-step tutorials for various scientific calculator models.
- **Coursera & Udemy:** Search for courses specifically designed for scientific calculator training. They often come with practical exercises.

3. Practice Worksheets

Practice is key to mastering the use of a scientific calculator. Look for worksheets that challenge you to solve problems using specific functions of the calculator.

Resources:
- **Math Websites**: Websites like Math-Aids.Com and Kuta Software offer customizable worksheets that can be solved using a scientific calculator.
- **Educational Platforms**: Platforms like Edmodo or Google Classroom may have resources shared by educators that specifically focus on using scientific calculators.

4. Join Forums and Communities

Engaging with communities can provide insights and solve doubts through peer learning.

Resources:
- **Reddit**: Subreddits like r/learnmath or r/calculators are great for asking questions and sharing knowledge.
- **Quora**: A platform where you can ask specific questions or read through existing threads on using scientific calculators.

5. Use Simulation Software

Before you invest in a high-end scientific calculator, or to practice without having the physical calculator with you, simulator software can be a handy resource.

Resources:
- **Virtual Calculators**: Websites like Desmos and GeoGebra offer free virtual scientific calculators.
- **Mobile Apps**: Look for apps that simulate the exact model of your scientific calculator. These can be extremely useful for on-the-go learning.

Tips and Tricks for Efficient Use

1. **Learn the Order of Operations**: Understanding how your calculator handles the order of operations (PEMDAS/BODMAS) can prevent common mistakes.
2. **Familiarize with Function Keys**: Knowing the shortcut keys and functions can save a lot of time during exams or when solving complex problems.
3. **Practice Regularly**: Incorporate the use of a scientific calculator into your daily study routine to become more proficient.
4. **Understand Error Messages**: Learning what different error messages mean can help you quickly troubleshoot and correct mistakes.
5. **Explore Beyond Basics**: As you become comfortable with basic operations, start exploring advanced functions like statistical analysis, calculus functions, and programming capabilities if your calculator supports them.

By leveraging these resources and tips, you can significantly improve your proficiency with a scientific calculator, making it a powerful tool in your academic and professional toolkit.

Practice Problems to Enhance Skills

Using a scientific calculator effectively can significantly enhance problem-solving skills in mathematics, physics, engineering, and other disciplines requiring complex calculations. Here's a detailed guide on how to leverage a scientific calculator to its fullest potential, complemented by practice problems to sharpen your skills.

Understanding the Basics

Before diving into complex functions, familiarize yourself with the basic operations of your scientific calculator, such as addition, subtraction, multiplication, division, and the use of parentheses to structure calculations properly.

Tip: Always check the order of operations on your calculator. Practice entering simple equations to see how it interprets operations.

Practice Problem: Calculate $(3 + 4) \times 5 - \frac{10}{2}$. This helps you understand how your calculator handles order of operations.

Exploring Scientific Functions

Scientific calculators come packed with functions for trigonometry, logarithms, exponents, and more. Explore these functions by referring to the calculator's manual and experiment with them.

Tip: Use the memory functions (M+, M-, MR, MC) to store intermediate results. This can simplify complex calculations that reuse certain values.

Practice Problem: Find the sine, cosine, and tangent of 45 degrees. Then, calculate the natural logarithm of the result. This familiarizes you with switching between trigonometric functions and logarithms.

Working with Complex Numbers

Handling complex numbers is straightforward with a scientific calculator that supports them. Learn how to switch to complex mode if your calculator requires it.

Tip: Practice converting between rectangular and polar forms of complex numbers using your calculator's functions, which is crucial for engineering and physics problems.

Practice Problem: Convert the complex number $(3 + 4i)$ to its polar form, and then find its absolute value and angle.

Solving Equations

Many scientific calculators offer equation-solving features. Dive into how to input equations and interpret the solutions provided by your calculator.

Tip: When solving equations, write down each step and check against your calculator's solution. This will help you catch any misunderstandings in how your calculator interprets inputs.

Practice Problem: Solve the quadratic equation $x^2 - 5x + 6 = 0$. Verify the roots using the calculator's equation solver and by manual calculation.

Graphing Functions

If your calculator has graphing capabilities, utilize this feature to understand functions better visually.

Tip: Start with simple functions to understand how to adjust viewing windows and interpret the graphs.

Practice Problem: Graph the function $y = x^2 - 4x + 4$. Identify the vertex and intercepts using the graphing tool.

Statistics and Probability

Scientific calculators often have functions to calculate mean, median, standard deviation, and probabilities. These can be incredibly useful in statistics.

Tip: Practice entering data sets into your calculator to compute these statistics. Check your manual for how to access and use these features.

Practice Problem: Calculate the mean and standard deviation of the dataset $[2, 3, 5, 7, 11]$. Then, explore any probability functions by calculating permutations or combinations of a set.

Continuous Learning

The key to mastering a scientific calculator is continuous exploration and practice. Challenge yourself with problems from different disciplines to discover new functions and shortcuts.

Tip: Regularly refer to the calculator manual and online resources. There are many tutorials and forums where users share tips and tricks.

Incorporating these practice problems and tips into your study routine can greatly enhance your proficiency with a scientific calculator. This not only speeds up calculations but also deepens your understanding of the underlying concepts in various scientific and mathematical domains.

Chapter: 9 Comparing Scientific Calculators

Features to Consider When Buying a Calculator

When purchasing a scientific calculator, several key features and considerations come into play, especially in relation to its use. Scientific calculators are indispensable tools for students, engineers, and professionals dealing with complex mathematical, scientific, and engineering calculations. Understanding the features to consider can significantly enhance your ability to use the calculator effectively. Here's a comprehensive guide:

1. Functionality and Type

- **Basic vs. Advanced Scientific Functions**: Ensure the calculator has basic scientific functions like trigonometric, logarithmic, and exponential functions. Advanced models include features like calculus operations, complex numbers, and matrix calculations. Depending on your needs, such as high school, college courses, or professional work, the required functionality may vary.

2. Usability

- **Ease of Use**: A user-friendly interface is crucial. Buttons should be well-labeled and responsive, with a logical layout for easy navigation.
- **Multi-line Display**: Look for calculators with multi-line displays that can show both the equation and the result simultaneously, facilitating easier review and editing of calculations.

3. Programmability

- **Programming Capability**: Some scientific calculators allow you to program them for complex or repetitive tasks, which can be a significant advantage for advanced mathematics or engineering.

4. Memory

- **Storage Memory**: Adequate memory allows you to store equations and data. Some calculators offer the ability to save and recall previous calculations, which is particularly useful during lengthy problem-solving sessions.

5. Connectivity

- **Connectivity Options**: Higher-end models may offer the ability to connect to computers or other devices, enabling the transfer of data or even software upgrades.

6. Display Quality

- **Screen Resolution and Type**: A high-resolution screen can make a big difference in readability, especially for graphs or when working with a large amount of data.

7. Power Source

- **Battery Life and Type**: Consider whether the calculator uses standard AAA batteries, a rechargeable battery, or solar power. A combination of solar and battery power offers reliability and convenience.

8. Durability and Portability

- **Build Quality**: A robust calculator that can withstand daily use and occasional drops is essential.
- **Size and Weight**: The calculator should be compact and lightweight enough for easy portability, without sacrificing usability or screen size.

How to Use a Scientific Calculator

Understanding the features is one thing, but knowing how to effectively use a scientific calculator is crucial:

- **Familiarize with the Manual**: Start by reading the manual to understand the basic and advanced functions.
- **Practice Basic Functions**: Get comfortable with common operations (addition, subtraction, multiplication, division) and gradually move to scientific functions (sin, cos, tan, log, ln, etc.).
- **Learn the Special Functions**: Understand how to use special functions such as converting between degrees and radians, using constants, and performing statistical calculations.
- **Explore Programming Features**: If your calculator has programming capabilities, learn how to create simple programs to automate repetitive tasks.
- **Utilize Memory Functions**: Learn to store and recall previous calculations or constants, which can save time.
- **Practice Graphing (if applicable)**: For calculators with graphing capabilities, practice plotting and analyzing functions.

Conclusion

Choosing the right scientific calculator involves balancing functionality, usability, and additional features against your specific needs and budget. Once you've selected a calculator, investing time to learn its features and how to use them effectively will maximize its utility in your academic or professional pursuits. Whether you're solving basic arithmetic problems or complex engineering equations, a solid understanding of both your calculator's capabilities and its operation is essential for success.

Pros and Cons of Popular Models

When it comes to harnessing the full potential of scientific calculators, understanding the strengths and limitations of popular models is crucial. Scientific calculators, designed for complex mathematics including trigonometry, statistics, algebra, calculus, and more, are indispensable tools for students, professionals, and enthusiasts. Let's delve into the pros and cons of some popular models in relation to their usage.

Texas Instruments TI-84 Plus

Pros:

- **User-friendly Interface**: The TI-84 Plus offers an intuitive interface that makes it easier for beginners to learn how to use a scientific calculator effectively.
- **Graphing Capabilities**: Its strong suit is the ability to graph multiple functions simultaneously, which is invaluable for calculus and algebra.
- **Programmable**: Users can write their own programs, which is a plus for those looking to perform repetitive tasks or custom calculations.

Cons:

- **Price**: It's on the pricier side, which might not be ideal for every budget.
- **Complexity for Beginners**: While it has many features, the sheer number of them can be overwhelming for beginners.

Casio FX-115ES Plus

Pros:

- **Natural Textbook Display**: This model displays equations as they appear in textbooks, making it easier to follow along and understand complex calculations.
- **Solar Power:** It includes a solar power function with a battery backup, ensuring that it's always ready for use.
- **Affordable**: It offers a high functionality-to-cost ratio, making it accessible to a wide range of users.

Cons:

- **Learning Curve**: Some users find the menu system less intuitive than its competitors, which can steepen the learning curve.
- **Limited Graphing**: It lacks the graphing capabilities of models like the TI-84 Plus, which could be a drawback for some advanced users.

HP 35s Scientific Calculator

Pros:

- RPN or Algebraic Entry: It allows users to choose between Reverse Polish Notation (RPN) and traditional algebraic entry, catering to a wide preference range.
- **Programmable**: Like the TI-84 Plus, it allows for custom programming, which can be a significant advantage for professionals and students.
- **Durability**: HP calculators are known for their build quality and reliability over time.

Cons:

- **Complex Interface**: The interface can be less intuitive for those accustomed to other brands, potentially requiring a longer adjustment period.
- Price: Similar to the TI-84 Plus, it's on the higher end of the price spectrum.

Usage Tips for Scientific Calculators

1. **Familiarize with the Manual**: Each model has unique features. Spending time with the manual can significantly speed up the learning process.

2. **Practice Regularly**: Regular practice can help you remember the functions and how to access them, making calculations faster and more efficient.
3. Utilize Online Resources: Many tutorials and forums are dedicated to specific models, which can be invaluable resources for troubleshooting and learning advanced techniques.
4. **Explore Beyond Basics**: Once comfortable with basic operations, exploring advanced functions can expand your capabilities and understanding of mathematical concepts.

Choosing the right scientific calculator depends on your specific needs, budget, and the complexity of the mathematics you intend to tackle. While high-end models offer extensive features, they may not be necessary for everyone. Assessing both the pros and cons in relation to your usage can lead to a more informed decision that aligns with your educational or professional requirements.

Recommendations Based on Use Case

When comparing scientific calculators and providing recommendations based on use cases, it's crucial to understand both the capabilities of these tools and the specific requirements of different users. Scientific calculators are indispensable in a wide range of academic and professional fields, offering functions beyond basic arithmetic, including trigonometry, calculus, statistics, and in some models, programmability and graphing capabilities. Here's a comprehensive guide to selecting a scientific calculator based on various use cases, coupled with insights on how to use these complex tools effectively.

High School Students

For high school students, a scientific calculator that offers basic functions like trigonometric calculations, logarithms, exponentials, and possibly some statistical functions is usually sufficient. The **Texas Instruments TI-30X IIS** and the **Casio FX-115ES PLUS** are excellent choices, providing ease of use, durability, and all the necessary functions for high school math and science courses. Learning to use these calculators involves familiarizing oneself with the function keys, understanding how to input expressions correctly, and utilizing the memory functions to store interim results.

College Students and Engineering Majors

College students, especially those in engineering or physical sciences, may require calculators with advanced features like matrix calculations, complex numbers, integration, differentiation, and possibly graphing capabilities. The **Texas Instruments TI-36X Pro** and the **Casio FX-991EX** are highly recommended for their advanced functions, which can handle the more complex calculations required in college-level courses. Users should delve into the manual to explore advanced functions, practice entering complex equations, and learn to interpret the output for analyses like integrals and matrix operations effectively.

Professionals and Researchers

Professionals in fields such as engineering, physics, and research who require a scientific calculator for work might need even more specialized features, including programmability, graphing functions, and the ability to connect to computers or other devices. The **HP 35s Scientific Calculator and** Texas Instruments TI-84 **Plus** are standout options, offering extensive functionality suitable for professional work. Mastery of these calculators involves understanding programming basics (for models that offer this feature), becoming proficient in graphing and analyzing data, and learning how to utilize connectivity options for data transfer and more comprehensive analyses.

Educators

Educators teaching mathematics or science might prefer scientific calculators that balance advanced functionality with ease of teaching. The **Casio FX-991EX** and **Texas Instruments TI-30X IIS** are both suitable, as they offer a wide range of functions and a user-friendly interface that can be easily demonstrated on a projector or through instructional materials. Educators should focus on learning how to navigate through functions quickly, understanding how to reset the calculator for classroom demonstrations, and exploring ways to integrate the calculator's functions into lesson plans effectively.

General Use

For those who need a scientific calculator for general use, including basic statistics, algebra, and geometry, user-friendliness and affordability might be the main concerns. The **Casio FX-300MS** and **Sharp EL-531XBGR** are ideal, providing essential scientific calculator functions in an easy-to-use format at a reasonable price. Users in this category should concentrate on familiarizing themselves with basic functions, understanding the order of operations, and using the calculator to check homework or solve everyday math problems.

How to Use a Scientific Calculator Effectively

1. **Read the Manual**: Each model has unique features. Understanding these can vastly improve your efficiency.
2. **Practice Common Functions**: Regular use of basic and complex functions will help you navigate the calculator more swiftly.
3. **Learn the Syntax**: Understand how your calculator interprets operations, especially for complex calculations.
4. **Use Memory Functions:** Learn how to store and recall numbers to streamline your calculations.
5. **Explore Advanced Features**: If your calculator has graphing or programmable functions, take time to learn these capabilities.

Selecting the right scientific calculator depends largely on the user's specific needs, whether academic, professional, or personal. By understanding the functions and learning how to use them effectively, users can significantly enhance their problem-solving skills and efficiency.

Chapter: 10 Ethical Considerations and Exam Policies

Using Calculators Responsibly

Using calculators, especially scientific ones, responsibly is an essential skill in both academic and professional settings. It's crucial to understand not only how to use these tools to solve complex problems but also to be aware of the ethical considerations and exam policies that govern their use.

Ethical Considerations

1. **Academic Integrity**: When using a calculator, it's vital to uphold the principles of honesty and integrity. This means acknowledging the assistance of the calculator in solving problems and understanding the underlying concepts instead of relying solely on the device for answers.

2. **Intellectual Development**: Calculators should serve as a supplement to learning, not a substitute. It's important to first attempt solving problems manually to grasp the fundamental concepts. Use calculators for

verification or to handle complex calculations that go beyond manual computation capabilities.

3. **Respect for Policies**: Many educational institutions have specific guidelines about calculator use during classes and exams. Abiding by these rules is a matter of respect and integrity. Unauthorized use can lead to consequences such as exam disqualification or disciplinary action.

Exam Policies

1. **Approved Models**: Institutions often specify which calculator models are allowed during exams to ensure fairness and prevent cheating through the use of advanced or programmable calculators. Always check the approved list before bringing a calculator to an exam.

2. **Clear Memory**: Some exams require students to clear their calculator's memory before starting to prevent the storage of notes or formulas that could give an unfair advantage. Knowing how to clear your calculator's memory is therefore crucial.

3. **Understanding Functions**: It's important to be familiar with the functions of your scientific calculator, as misuse can lead to incorrect answers. For instance, knowing how to use the trigonometric, logarithmic, and

statistical functions accurately is crucial for subjects like mathematics, physics, and engineering.

4. **Practice**: Before an exam, practice with the same model of calculator you'll use on test day. This helps in becoming efficient with its functions and reduces the risk of mistakes under exam conditions.

5. **Calculator Etiquette**: During exams, use your calculator quietly and respectfully to avoid disturbing others. The sound of clicking buttons can be distracting in a silent exam hall.

Conclusion

Using a scientific calculator responsibly involves more than just knowing how to operate the device. It encompasses understanding and respecting the ethical implications and adhering to the specific policies set by educational or professional institutions. By using calculators judiciously, students and professionals not only ensure fairness and integrity but also reinforce their understanding of the subject matter, ultimately leading to a more meaningful and enriched learning experience.

Understanding Examination Rules Regarding Calculators

Understanding examination rules regarding the use of calculators, especially scientific calculators, is crucial for students aiming for academic success. As technology advances, scientific calculators have become more sophisticated, offering functionalities that can significantly aid in complex problem-solving. However, these capabilities also raise concerns about fairness and integrity during examinations. Here, we explore the ethical considerations and exam policies related to the use of scientific calculators, guiding students on how to ethically use these tools in an exam setting.

Ethical Considerations

1. **Fairness**: The primary ethical consideration is ensuring fairness among all exam takers. Access to advanced calculators should not give any student an undue advantage over others. Ethically, students must use only the functionalities allowed by the examination guidelines.

2. **Integrity**: Students must uphold integrity by using calculators in a manner that reflects their understanding and abilities. Relying on calculators for tasks they are supposed to perform manually (e.g., basic arithmetic)

can undermine their learning process and academic integrity.

3. **Respect for Guidelines**: Adhering to exam rules is a matter of respect for the examination process and the institutions that administer them. Violating these rules not only disrespects the process but also jeopardizes the student's academic record.

Exam Policies

Exam policies on calculator use vary by institution and the level of study. However, some common policies include:

1. **Approved Models**: Many exams specify models or types of calculators that are permitted. For instance, some may allow basic scientific calculators but not graphing calculators or calculators capable of symbolic manipulation.

2. **Memory Clearance**: To prevent storing and retrieving notes, some exams require students to clear their calculator's memory before the exam begins. Proctors may also inspect calculators to ensure compliance.

3. **Functionality Limitations**: Certain exams may restrict the use of calculators to specific functions. For

example, calculators that can perform symbolic algebra or have embedded formulas may be banned.

4. **Physical Inspections**: Calculators may be subject to physical inspection to check for unauthorized modifications or hidden notes.

How to Use a Scientific Calculator Ethically in Exams

1. **Familiarize with Allowed Features**: Before the exam, understand which calculator features are permitted and practice using them. This includes knowing how to perform complex calculations that are relevant to the exam material.

2. **Prepare the Calculator**: Ensure your calculator is in compliance with the exam policies. This may involve clearing the memory, removing any stored data, or even resetting the device.

3. **Practice Ethical Usage**: Use the calculator as a tool to aid your problem-solving, not as a crutch. Practice doing calculations manually when possible, and use the calculator for checking work or performing complex operations.

4. **Understand Manual Calculations**: While calculators are helpful, it's important to understand the underlying

principles of the calculations you're performing. This ensures you're not solely reliant on the device and can work through problems if the calculator fails.

5. **Avoid Sharing During Exams**: Sharing calculators during exams can lead to misunderstandings or accusations of cheating. Bring your own approved device and ensure it's prepared according to the exam guidelines.

In conclusion, the ethical use of scientific calculators in examinations revolves around fairness, integrity, and adherence to established guidelines. By understanding and respecting these principles, students can effectively and ethically leverage their calculators as tools for academic success. It's not just about knowing how to use a scientific calculator, but understanding when and where its use is appropriate and in alignment with the spirit of academic honesty.

Conclusion

Understanding how to use a scientific calculator effectively culminates in the ability to reach accurate conclusions in mathematical and scientific problems. The essence of mastering a scientific calculator extends beyond the mere input of numbers and operations; it encompasses a deep comprehension of the problem at hand, the logical sequence of operations required to solve it, and the interpretation of the results provided by the calculator.

In the realm of scientific calculations, the precision and reliability of conclusions drawn are paramount. A scientific calculator, equipped with functions that range from basic arithmetic to complex equations, trigonometry, and statistical analysis, serves as a potent tool in achieving this precision. However, the tool's efficacy is significantly dependent on the user's familiarity with its functions and the underlying principles of the calculations being performed. This familiarity ensures that the calculator is used to its fullest potential, not as a crutch but as an enhancer of the user's mathematical prowess.

To utilize a scientific calculator most effectively, one must first thoroughly understand the mathematical concepts before applying them. This foundational knowledge enables the user to select the appropriate functions on the calculator and to sequence operations correctly. Misapplication or misunderstanding of these functions can lead to errors in calculation, leading to inaccurate conclusions. Hence, a dual focus on conceptual understanding and operational proficiency with the calculator is essential.

Moreover, interpreting the results correctly is crucial in drawing accurate conclusions. The calculator might provide a result in a form that requires further interpretation or conversion, especially in scientific notation or when dealing with units of measurement. Users must be adept at making these conversions and interpreting these results in the context of the problem.

Another important aspect is the recognition of the limitations of the calculator. While scientific calculators can perform a vast array of operations, they are not infallible. Understanding the limitations of the calculator's algorithms, especially when dealing with extremely large numbers, approximations, or highly complex equations, is crucial. This awareness aids in critically evaluating the results provided by the calculator, ensuring that the conclusions drawn are not blindly accepted but are scrutinized for accuracy.

In essence, drawing accurate conclusions using a scientific calculator involves more than just pressing buttons; it requires a deep engagement with the mathematical concepts, a strategic approach to using the calculator's functions, and a critical eye for interpreting the results. Mastery of these elements ensures that the scientific calculator becomes a valuable ally in exploring and understanding the complex world of mathematics and science, empowering users to reach precise and reliable conclusions.

www.ingramcontent.com/pod-product-compliance
Lightning Source LLC
Chambersburg PA
CBHW071054240526
45471CB00015B/1875